Bautechnisches Englisch im Bild
Illustrated Technical German for Builders

W. K. Killer

Bautechnisches Englisch im Bild

Illustrated Technical German for Builders

5., durchgesehene Auflage

4th reprint with amendements

BAUVERLAG GMBH · WIESBADEN UND BERLIN

CIP-Kurztitelaufnahme der Deutschen Bibliothek
Killer, Wilhelm K.:
Bautechnisches Englisch im Bild = Illustrated
technical German for builders / W. K. Killer. —
5., durchges. Aufl. — London : Godwin ; Wiesbaden ;
Berlin : Bauverlag, 1981.

ISBN 3-7625-1477-1 (Bauverlag)

Das Werk ist urheberrechtlich geschützt.
Die dadurch begründeten Rechte, insbesondere die der Übersetzung,
des Nachdruckes, der Entnahme von Abbildungen, der Funksendung,
der Wiedergabe auf photomechanischem oder ähnlichem Wege
(Fotokopie, Mikrokopie) und der Speicherung in Datenverarbeitungs-
anlagen, bleiben, auch bei nur auszugsweiser Verwertung, vorbehalten.

1. Auflage 1973 1st edition
2. Auflage 1974 1st reprint
3. Auflage 1976 2nd reprint
4. Auflage 1977 3rd reprint
5. Auflage 1981 4th reprint
Nachdruck der 5. Auflage 1984 5th reprint
© 1981 by Bauverlag GmbH, Wiesbaden und Berlin
Druck: Druckhaus Hans Meister KG, Kassel
ISBN 3-7625-1477-1

INHALT / CONTENT

	Seite/Page
Architektur Architecture	3
Hoch- und Tiefbau (Bauingenieurwesen) Building and Civil Engineering	19
Baukunde Building Knowledge	31
Baustelle Building Site	47
Baumaschinen Building Machinery	49
Bauteile Structural Elements	51
Unterbau Substructure	53
Oberbau Superstructure	60
Betonarbeiten Concrete Work	69
Bewehrung Reinforcement	76
Schalarbeiten Formwork	80
Maurerarbeiten Brickwork	83
Stahl- und Metallarbeiten Steel- and Metalwork	97
Holzarbeiten Timberwork	111
Dächer, Installationen und Ausbau Roofs, Plumbing and Finishes	137
Alphabetisches Verzeichnis der bautechnischen Begriffe Alphabetical Index of Technical Terms in Construction	169

Vorwort

Mit diesem Bildwörterbuch gibt der Autor, selbst jahrelang im englischsprachigen Ausland tätig, ein Hilfsmittel in die Hand, das die englische Baufachsprache in anschaulicher Weise erschließt. Detaillierte Zeichnungen aus allen Bereichen des Bauwesens, darunter aus zahlreichen Teilgebieten wie Fliesenarbeiten, Schmiedearbeiten, Fensterbau, Elektro-Installation, Schalarbeiten u.a.m. illustrieren sowohl allgemeinere Begriffe als auch Spezialausdrücke, wobei besondere Formulierungen zusätzlich aufgeführt werden. Die Darstellungen sind jeweils mit den englischen und deutschen Begriffen kombiniert.
Das Buch ist nach folgenden Sachgebieten gegliedert: Architektur, Bauingenieurwesen, Baukunde, Baustelle und Baumaschine, Unterbau, Oberbau, Betonarbeiten, Bewehrung, Schalarbeiten, Maurerarbeiten, Stahl- und Metallarbeiten, Holzarbeiten, Dächer, Installationen und Ausbau.
Ein zusätzliches Arbeitsmittel ist das alphabetische Wörterverzeichnis, in dem ca. 1.650 deutsche Baufachbegriffe mit Hinweisen auf die betreffende Seitenzahl enthalten sind. Man kann also mit diesem Bildwörterbuch leicht das notwendige „Bauenglisch" erlernen und „Bautechnisches Englisch im Bild" gleichzeitig zum Nachschlagen verschiedener Fachtermini des Bauwesens benutzen.

Preface

With this illustrated dictionary the author, who has been working for many years in English-speaking countries, has given us a manual that explains pictorially the technical language of the English and German building trade. Detailed drawings of all aspects of building, among them numerous sub-divisions such as tile-work, ironwork, window construction, electrical installation, form-work, etc., illustrate general concepts as well as technical terms and wherever specialised expressions exist these are also added. In the illustrations the terms are given in both English and German.
The book covers the following subjects: architecture, civil engineering, building theory, the building site and building machinery, substructure and superstructure, concrete work, reinforcement, form-work, masonry, steel and metal-work, timberwork, roofing, installation work and finishing.
Of further assistance is the alphabetical glossary which contains. c. 1,650 technical terms in German with the relevant page references. From this illustrated dictionary therefore, one can easily learn the necessary "English building-language" and at the same time use "Illustrated Technical German for Builders" as a reference for the various professional terms in the building industry.

ILLUSTRATED TECHNICAL GERMAN FOR BUILDERS

BAUTECHNISCHES ENGLISCH IM BILD

Architecture
Architektur

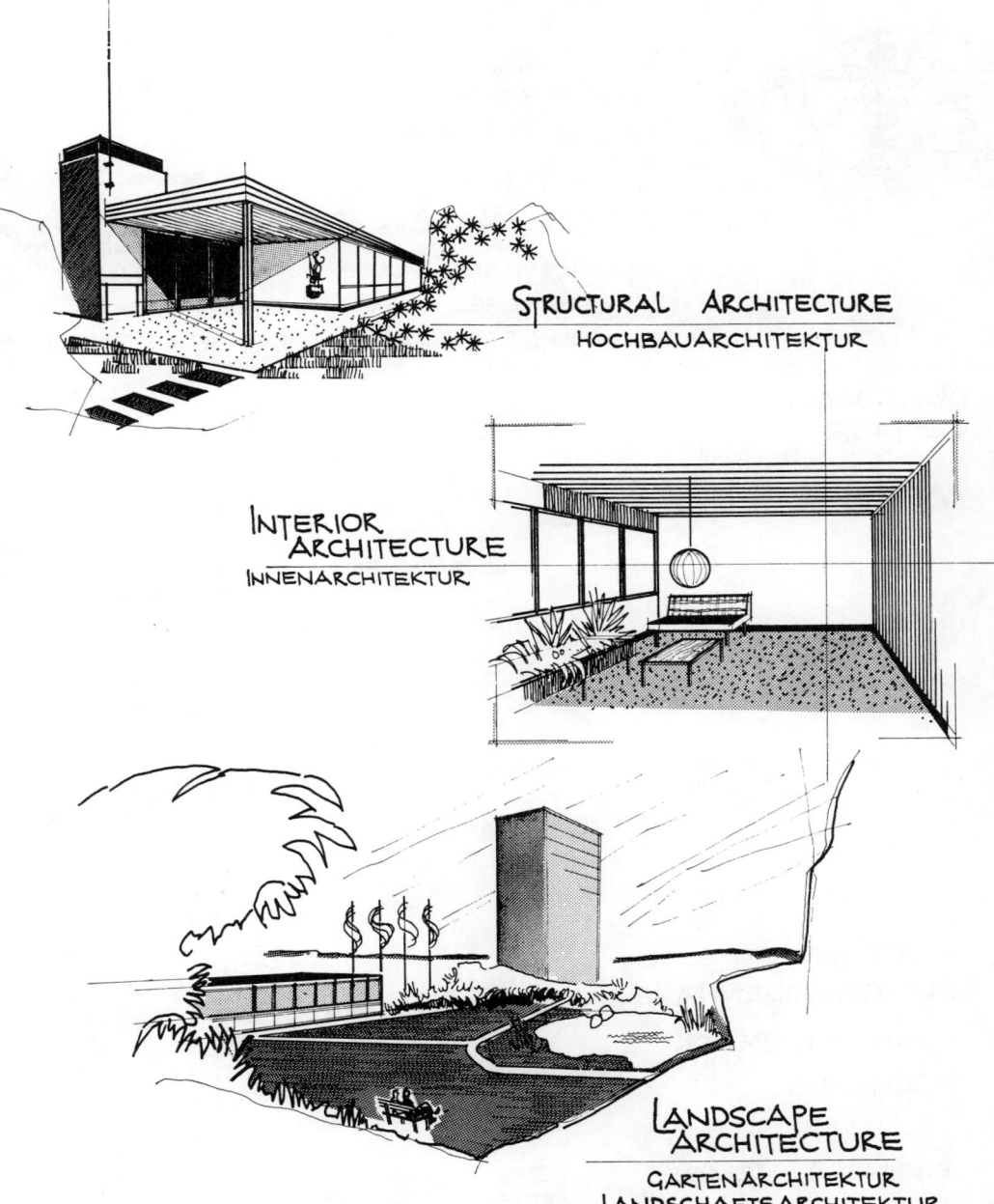

Structural Architecture
Hochbauarchitektur

Interior Architecture
Innenarchitektur

Landscape Architecture
Gartenarchitektur
Landschaftsarchitektur

Environmental Architecture
Landschaftsgebundene Architektur

ARCHITEKTUR   ARCHITECTURE

SOME TYPES OF BUILDINGS
EINIGE ARTEN VON BAUWERKEN

HOUSE HOME DWELLING	WOHNHAUS EIGENHEIM

EXAMPLES: BEISPIELE:

SPLIT-LEVEL HOUSE   WOHNHAUS MIT VERSETZ-
                             TEN GESCHOSSEBENEN
STEPPED HILLSIDE HOUSE   TERRASSENHAUS
SEMI-DETACHED HOME   ZWEISPÄNNERHAUS
TERRACE-HOUSE DEVELOPMENT   REIHENHAUSSIEDLUNG

BUNGALOW
COTTAGE   SOMMERHAUS, FERIENHAUS

FLAT
APARTMENT   WOHNUNG

FOR EXAMPLE: BACHELOR FLAT   JUNGGESELLENWOHNUNG
                              APARTMENT
                              EINZIMMERWOHNUNG

SHACK, SHANTY, SHED, HUT ——— HÜTTE, BUDE

RECREATION CENTRE   SPORTZENTRUM
                           ERHOLUNGSZENTRUM
INDOOR SWIMMING POOL   HALLENBAD

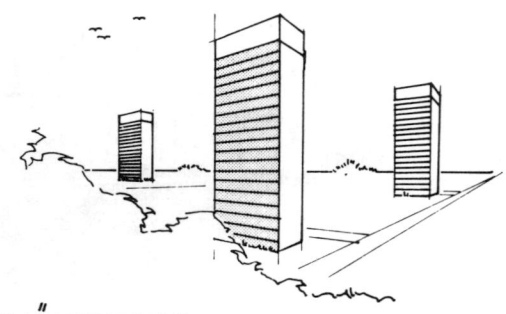

MULTI-STOREY BLOCK
MEHRFAMILIENHAUS

HIGH-RISE BLOCK
HOCHHAUS

BUSINESS PREMISES   GESCHÄFTSHAUS
                           GESCHÄFTSRÄUME

ARCHITEKTUR   ARCHITECTURE

AN ARCHITECT RENDERS THE FOLLOWING SERVICES:
  EIN ARCHITEKT ERBRINGT FOLGENDE DIENSTLEISTUNGEN:

DESIGN   ENTWURF, PLANUNG

   PRELIMINARY DESIGN   VORENTWURF
   FINAL DESIGN   ENDGÜLTIGER ENTWURF

SKETCH   SKIZZE

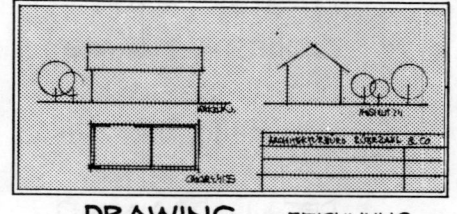
DRAWING   ZEICHNUNG

NOTE: **DESIGN** OF A REINFORCED CONCRETE BEAM
       (STATISCHE) **BEMESSUNG** EINES STAHLBETONBALKENS

SUPERVISION   ÜBERWACHUNG, BAULEITUNG

   SITE SUPERVISION   ÖRTLICHE BAULEITUNG
   PROJECT SUPERVISION   PROJEKTÜBERWACHUNG

"QUANTITY SURVEYORS" (MASSENBERECHNER) ERHALTEN IN ENGLISCH-
SPRACHIGEN LÄNDERN EINE GESONDERTE AUSBILDUNG UND BILDEN
EINE EIGENE BERUFSGRUPPE

A QUANTITY SURVEYOR'S SERVICES ARE:
  DIE DIENSTLEISTUNGEN EINES MASSENBERECHNERS SIND:

   ESTIMATES   KOSTENSCHÄTZUNGEN

   DETERMINATION OF QUANTITIES
                                MASSENERMITTLUNG

   PREPARATION OF BILLS OF QUANTITIES
                ERSTELLEN VON MASSEN-LEISTUNGSVERZEICHNISSEN

   REQUEST FOR TENDERS (QUOTATIONS, BIDS)
                EINHOLUNG VON ANGEBOTEN

   AWARD OF CONTRACTS
                VERGABE VON AUFTRÄGEN (VERTRÄGEN)

   ACCOUNTING   ABRECHNUNG

ARCHITEKTUR   ARCHITECTURE

List of PLANNING WORK performed by
The Architect

PRELIMINARY SKETCH	VORLÄUFIGE SKIZZE
PRELIMINARY DESIGN	VORENTWURF
FINAL DESIGN	ENDGÜLTIGER ENTWURF
KEY PLAN, BLOCK PLAN	ÜBERSICHTSPLAN
MASTER PLAN	HAUPTBEBAUUNGSPLAN
GENERAL LAYOUT	GENERALPLAN
DETAIL DESIGN	DETAILENTWURF
WORKING DRAWING	AUSFÜHRUNGSZEICHNUNG, WERKPLAN
DETAIL DRAWING	DETAILPLAN
RECESS DRAWING	AUSSPARUNGSPLAN
SITE SUPERVISION	ÖRTLICHE BAULEITUNG
FINAL APPROVAL	BAUABNAHME
TIME SCHEDULE / BUILDING PROGRAMME	BAUZEITPLAN

The Engineer

STRUCTURAL CALCULATION	STATISCHE BERECHNUNG
FORMWORK DRAWING	SCHALPLAN
REINFORCEMENT DRAWING	BEWEHRUNGSPLAN

The Quantity Surveyor

BILLS OF QUANTITIES	LEISTUNGSVERZEICHNISSE
CALLING FOR TENDERS	ANGEBOTSEINHOLUNG
CONTRACTS	VERTRAGSABSCHLÜSSE
PAYMENT SCHEDULE	FINANZIERUNGSPLAN

The Consultant (Architect or Engineer)

FEASIBILITY STUDY	DURCHFÜHRBARKEITSSTUDIE
EXPERT'S OPINION	GUTACHTEN
" SURVEY	"
" REPORT	"

ARCHITEKTUR   ARCHITECTURE

ARCHITECTURAL TERMS
ARCHITEKTONISCHE FACHAUSDRÜCKE

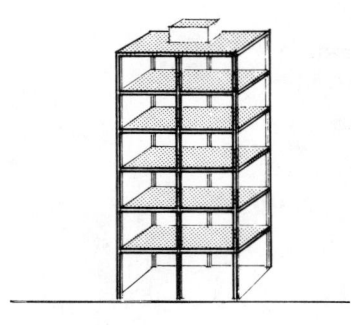 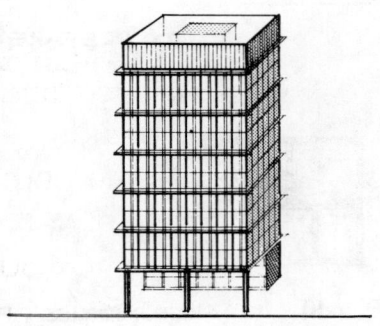

| STRUCTURE | FINISHED BUILDING |
| ROHBAU | FERTIGER BAU |

FINISH   AUSBAU

COVERAGE   ÜBERBAUBARE FLÄCHE
COVERED AREA   ÜBERBAUTE FLÄCHE, ÜBERDACHTE ~
CUBAGE, CUBICAL CONTENT   UMBAUTER RAUM, RAUMINHALT
LIVING AREA   WOHNFLÄCHE
COMMERCIAL AREA   GEWERBLICHE NUTZFLÄCHE

RESIDENTIAL AREA   WOHNGEBIET
HOUSING PROJECT }   WOHNSIEDLUNGSPLAN
HOUSING SCHEME
HOUSING SETTLEMENT }   WOHNSIEDLUNG
HOUSING ESTATE
TURN-KEY PROJECT   SCHLÜSSELFERTIGES PROJEKT

INDUSTRIAL AREA   INDUSTRIEGEBIET
INDUSTRIAL DEVELOPMENT AREA   INDUSTRIEANSIEDLUNGSZONE

BUILDING COSTS   BAUKOSTEN
PROGRESS PAYMENT   ABSCHLAGSZAHLUNG
FINAL PAYMENT   ENDZAHLUNG

| ARCHITEKTUR | ARCHITECTURE |

DRAWING   PLAN, ZEICHNUNG

TO PREPARE A DRAWING
EINEN PLAN ANFERTIGEN (MACHEN)

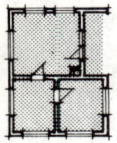

PLAN   GRUNDRISS, PLAN

PART PLAN   TEILGRUNDRISS
KEY PLAN   ÜBERSICHTSPLAN

LAYOUT
LAGEPLAN, BEBAUUNG
GESAMTGRUNDRISS

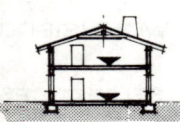

ELEVATION
(KONSTRUIERTE) ANSICHT

NORTH ELEVATION
SOUTH ~
EAST ~
WEST ~

CROSS SECTION   LONGITUDINAL SECTION
TRANSVERSAL SECTION   LÄNGSSCHNITT
QUERSCHNITT

ARCHITEKTUR   ARCHITECTURE

VIEW   ANSICHT (NICHT FRONTAL, SCHRÄG)
PANORAMIC VIEW

BIRD'S-EYE VIEW
VOGELPERSPEKTIVE

ISOMETRIC VIEW
ISOMETRISCHE ANSICHT

WORM'S-EYE VIEW
FROSCHPERSPEKTIVE

PERSPECTIVE VIEWS
PERSPEKTIVEN

CAPTION OF A DRAWING
KOPF EINER ZEICHNUNG

SUBJECT : GEGENSTAND :	DRAWING NUMBER : ZEICHNUNGSNUMMER :
	SCALE : MASSTAB :
	DATE : DATUM :
PROJECT : PROJEKT :	DRAWN BY : GEZEICHNET :
	CHECKED BY : GEPRÜFT :
	APPROVED BY : GENEHMIGT :

REVISION   ÜBERARBEITUNG

CORRECTION   KORREKTUR

ARCHITEKTUR COMMON IMPERIAL SCALES ARCHITECTURE
GEBRÄUCHLICHE MASSTÄBE IM ENGLISCHEN MASSYSTEM

1" = 1 INCH
1'-0" = 1 FOOT

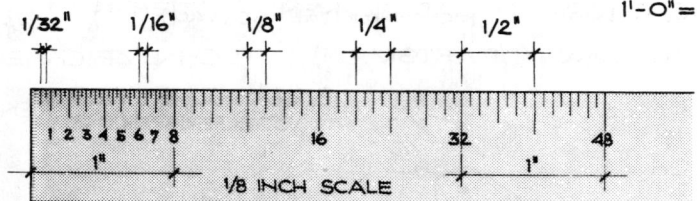

INCH SCALES

DETAIL & WORKING DRAWINGS
DETAIL- UND WERKPLÄNE
(AUSFÜHRUNGS-ZEICHNUNGEN)

F.S. (FULL SIZE)	=	1 : 1
1/2 F.S.	=	1 : 2
1/4 F.S.	=	1 : 4
1/2" = 1'-0"	=	1 : 24
1/4" = 1'-0"	=	1 : 48
1/8" = 1'-0"	=	1 : 96

SCHEME DESIGNS & LAND SURVEYORS DRAWINGS
ÜBERBAUUNGS- UND VERMESSUNGSPLÄNE

1/16" = 1'-0"	=	1 : 192
1/2" = 10'-0"	=	1 : 240
1/4" = 10'-0"	=	1 : 480
1/8" = 10'-0"	=	1 : 960
1/16" = 10'-0"	=	1 : 1920

N.T.S. (NOT TO SCALE) = O.M. (OHNE MASSTAB)

PARTS OF AN INCH ARE USUALLY GIVEN IN FRACTIONS
TEILE EINES ZOLLS WERDEN GEWÖHNLICH IN BRUCHTEILEN ANGEGEBEN

1/4"   NOT   ~~0,25"~~

USE THESE FRACTIONS :   BENÜTZE :

1/2"   1/4"   1/8"   1/16"   1/32"   1/64"

AVOID :   VERMEIDE :

1/3"   1/5"   1/6"   1/7"   1/9"

ARCHITEKTUR   ARCHITECTURE

    TO   DRAW   ZEICHNEN

    TO   DRAUGHT (DRAFT)   ABZEICHNEN, AUSZIEHEN

            DRAUGHTSMAN (DRAFTSMAN)   TECHN. ZEICHNER

    TO   TRACE   DURCHZEICHNEN, DURCHDRÜCKEN

    TO   SKETCH   SKIZZIEREN

LEGEND :   LEGENDE

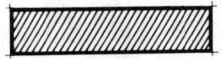   

HATCHED   SMUDGED
SCHRAFFIERT   GESCHUMMERT

   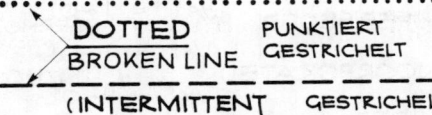

(INTERMITTENT   GESTRICHELT)
UNGEBRÄUCHLICH

REVISION   ÜBERARBEITUNG   PROPOSAL   VORSCHLAG
                                    PROPOSED LAYOUT
ALTERATION   ÄNDERUNG   GEPLANTE BEBAUUNG

SURVEY   AUFMASS   REFERENCE   BEZUG
               MASSAUFNAHME
                                    REFERENCE DRAWING
                                    BEZUGSZEICHNUNG
                                    GRUNDLAGENZEICHNUNG

NOTES ON DRAWINGS: THIS DRAWING IS TO BE READ
HINWEISE AUF ZEICHNUNGEN: IN CONJUNCTION WITH DRG.NO. ....
                          DIESE ZEICHNUNG IST IM ZUSAMMENHANG
                          MIT ZEICHG.NR. ...... ZU LESEN

ALL MEASURES (DIMENSIONS) ARE TO BE TAKEN (CHECKED) ON SITE
     ALLE MASSE SIND AM BAU ZU NEHMEN (ZU ÜBERPRÜFEN)

                          ANY QUERIES TO BE CLARIFIED
                          WITH THE ARCHITECT
                          UNSTIMMIGKEITEN SIND MIT DEM ARCHITEKTEN
                                    ZU KLÄREN

ARCHITEKTUR   ARCHITECTURE

GRID   GRID LINE
RASTER   RASTERLINIE

2,0 m   ACHSMASS / ACHSABSTAND
6'-7 7/16"   CC'S (CENTERS)

₵ = CENTRE LINE   ACHSLINIE

VARIOUS REVISIONS :
VERSCHIEDENE ÜBERARBEITUNGEN :

    REVISED   REVIDIERT, ÜBERARBEITET
    CHANGED   VERÄNDERT
    ADJUSTED   BERICHTIGT
    DELETED   GESTRICHEN
    ALTERED   GEÄNDERT
    SUPERSEDED BY ....   ÜBERHOLT DURCH .....
    INCORPORATED   HINZUGENOMMEN
    ADDED   HINZUGEFÜGT
    INTRODUCED   EINGEFÜHRT, EINGEZOGEN
    LOCATED   MASSLICH FIXIERT

TRUE NORTH / NORDEN   W · E · S

DRAWING PAPERS :
ZEICHENPAPIERE

                 TRACING PAPER
                   TRANSPARENT PAPIER
     CARTRIDGE PAPER OR BRISTOL BOARD
                     KARTONPAPIER

COPIES OF DRAWINGS :
ZEICHNUNGSKOPIEN

             SEPIA PRINT   MUTTERPAUSE
             PAPER PRINT   PAPIERPAUSE
             PLASTIC PRINT   PLASTIKPAUSE
             BLUE PRINT   BLAUPAUSE

ARCHITEKTUR ARCHITECTURE

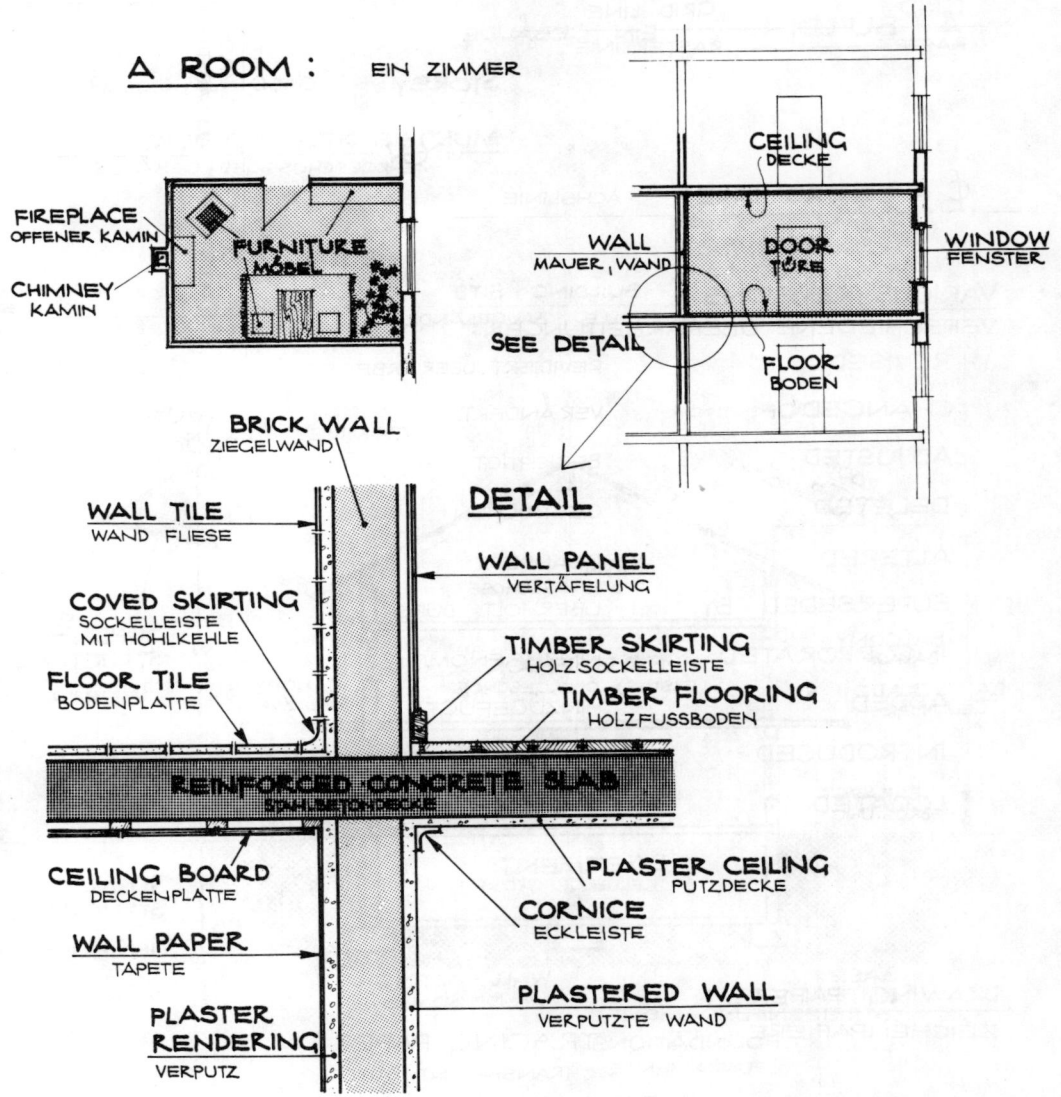

ARCHITEKTUR — ARCHITECTURE

EXAMPLES:

A BUILDING : EIN GEBÄUDE

STOREY — GESCHOSS, STOCKWERK

MULTI-STOREY BUILDING — MEHRGESCHOSSIGES GEBÄUDE

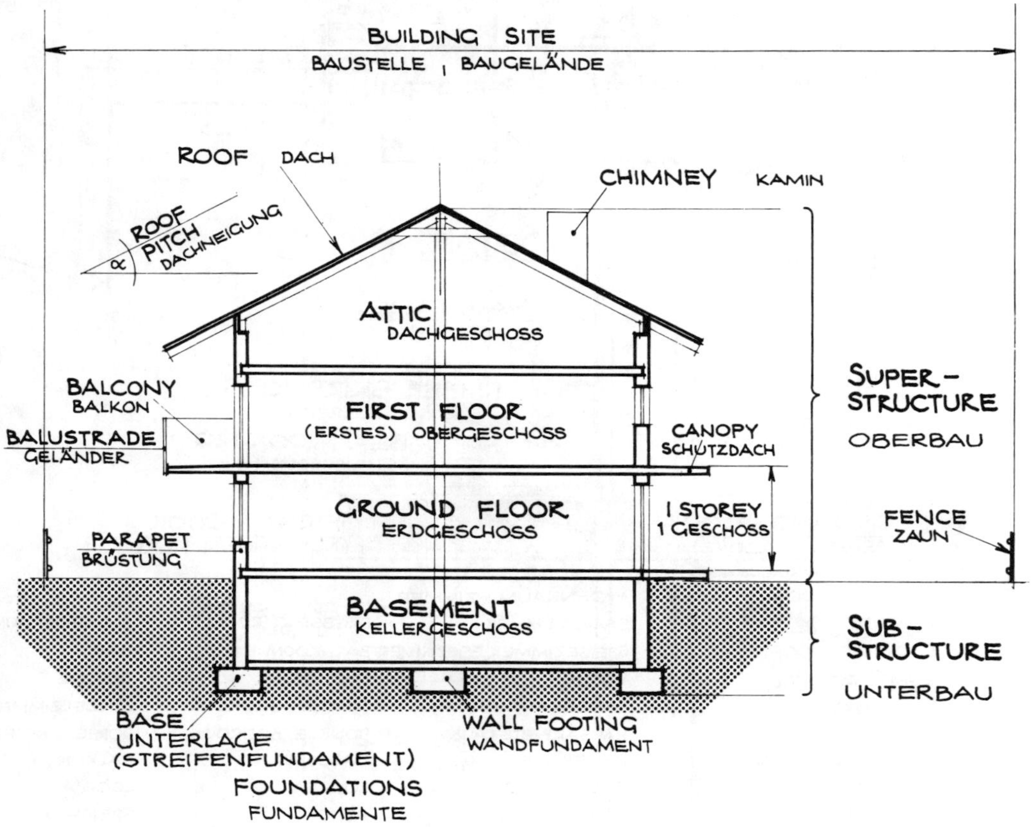

- BUILDING SITE — BAUSTELLE, BAUGELÄNDE
- ROOF — DACH
- ROOF PITCH — DACHNEIGUNG
- CHIMNEY — KAMIN
- ATTIC — DACHGESCHOSS
- BALCONY — BALKON
- BALUSTRADE — GELÄNDER
- FIRST FLOOR — (ERSTES) OBERGESCHOSS
- CANOPY — SCHUTZDACH
- SUPER-STRUCTURE — OBERBAU
- PARAPET — BRÜSTUNG
- GROUND FLOOR — ERDGESCHOSS
- 1 STOREY — 1 GESCHOSS
- FENCE — ZAUN
- BASEMENT — KELLERGESCHOSS
- SUB-STRUCTURE — UNTERBAU
- BASE — UNTERLAGE (STREIFENFUNDAMENT)
- WALL FOOTING — WANDFUNDAMENT
- FOUNDATIONS — FUNDAMENTE
- SUB-BASEMENT — UNTERES KELLERGESCHOSS

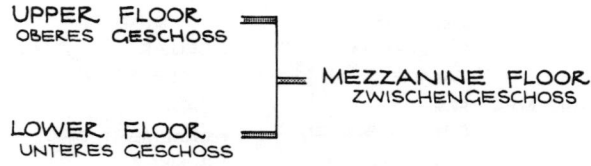

- UPPER FLOOR — OBERES GESCHOSS
- MEZZANINE FLOOR — ZWISCHENGESCHOSS
- LOWER FLOOR — UNTERES GESCHOSS

ARCHITEKTUR — ARCHITECTURE

EXAMPLES:

A PRIVATE HOME
EIN PRIVATHAUS

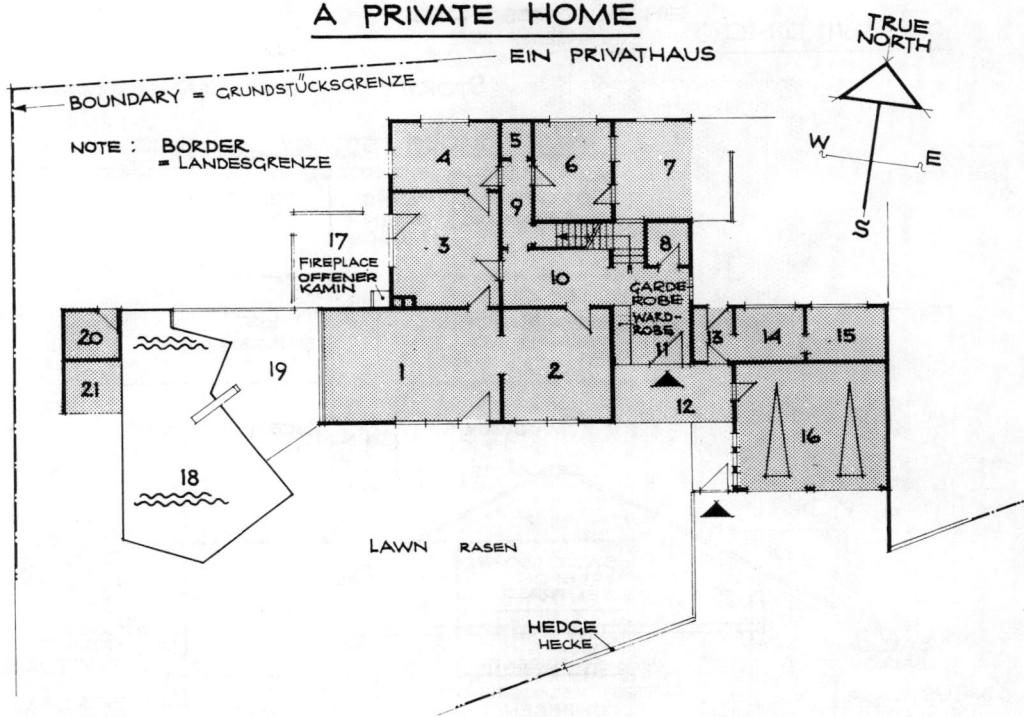

BOUNDARY = GRUNDSTÜCKSGRENZE
NOTE: BORDER = LANDESGRENZE

LEGEND:

GROUND FLOOR
(SHOWN ABOVE)

1	LIVING ROOM, ~ AREA	WOHNRAUM, -ZIMMER
2	STUDY	STUDIERZIMMER
3	DINING ROOM	SPEISEZIMMER, ESSZIMMER
4	KITCHEN	KÜCHE
5	PANTRY	SPEISEKAMMER
6	LAUNDRY	HAUSARBEITSRAUM
7	BACK YARD, PATIO	HINTERHOF
8	TOILET, WC	TOILETTE, WC
9	PASSAGE	FLUR, GANG
10	HALL	DIELE
11	ENTRANCE	EINGANG
12	COVERED AREA	ÜBERDACHTE FLÄCHE
13	PASSAGE	DURCHGANG
14	WORKSHOP	WERKSTATT
15	UTILITY ROOM	ZÄHLERRAUM
16	CARPORT	GARAGE
17	PATIO, OUTDOOR LIVING AREA	FREISITZ
18	SWIMMING POOL	SCHWIMMBECKEN
19	TERRACE	TERRASSE
20	UTENSIL ROOM	GERÄTERAUM
21	ALCOVE	ALKOVEN

FIRST FLOOR
(NOT SHOWN HERE)

DRESSING ROOM	ANKLEIDERAUM
BATHROOM	BAD
SHOWER	DUSCHE
SINGLE BEDROOM	EINZELZIMMER
DOUBLE BEDROOM	DOPPELZIMMER
BALCONY	BALKON
LOGGIA	LOGGIA
ATTIC	SPEICHER

BASEMENT
(NOT SHOWN HERE)

CELLAR	KELLER
STORE ROOM	VORRATSRAUM
SAUNA	SAUNA
BASEMENT BAR	KELLERBAR
BUNKER	LUFTSCHUTZRAUM

ARCHITEKTUR   ARCHITECTURE

EXAMPLES: **AN OFFICE FLOOR**
EIN BÜROGESCHOSS

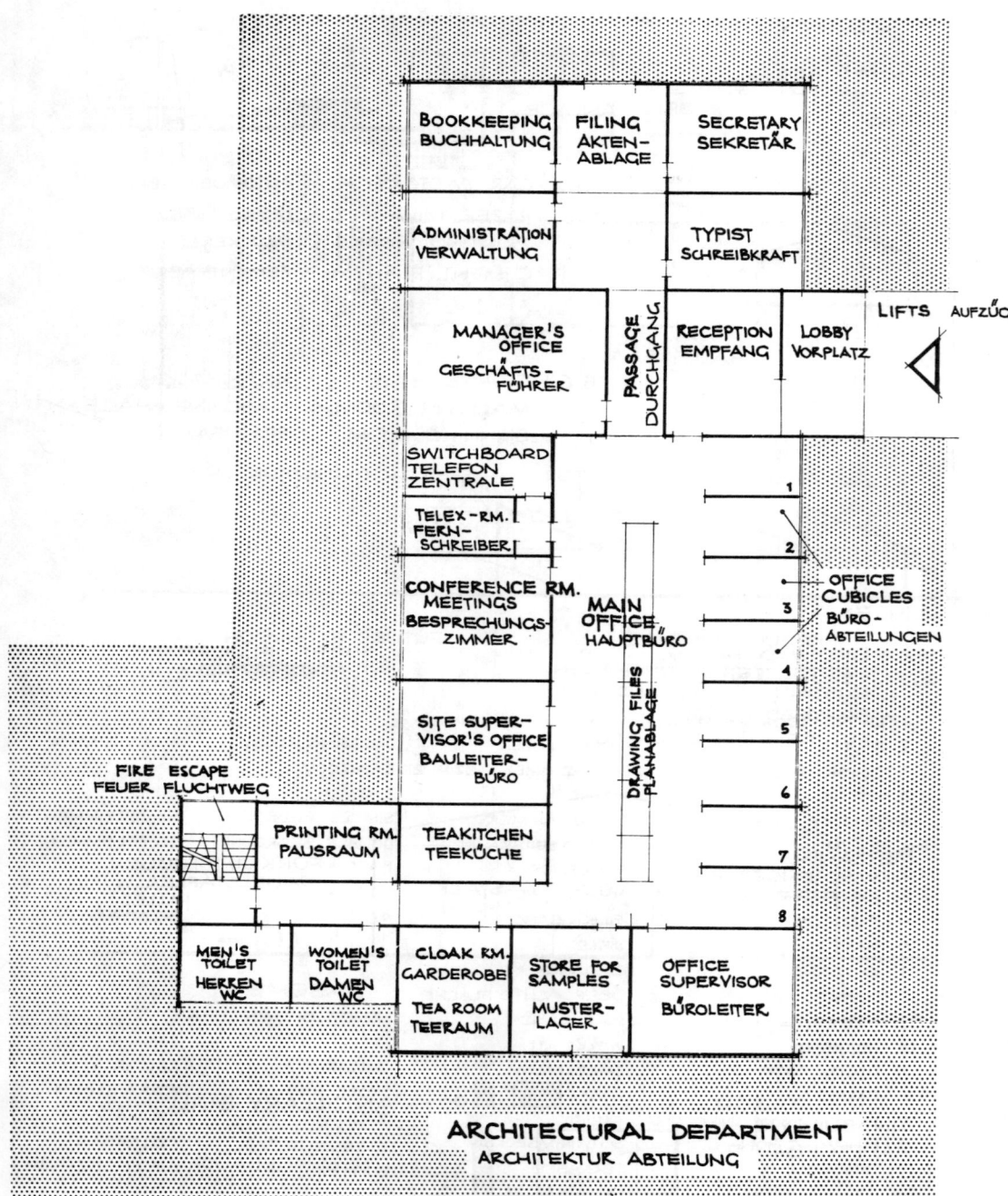

ARCHITECTURAL DEPARTMENT
ARCHITEKTUR ABTEILUNG

ARCHITEKTUR   ARCHITECTURE

EXAMPLES:
INDUSTRIAL BUILDING
INDUSTRIEBAU

FACTORY ⎫
MILL   ⎬ FABRIK, ANLAGE
WORKS   ⎪
PLANT   ⎭

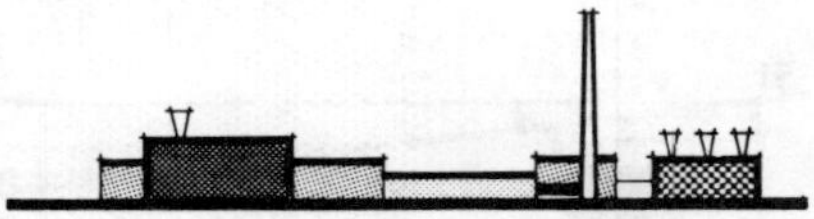

E.G. CAR FACTORY   AUTOMOBILFABRIK
      PAPER MILL   PAPIERFABRIK
      PRINTING WORKS   DRUCKEREI
      CEMENT PLANT   ZEMENTFABRIK

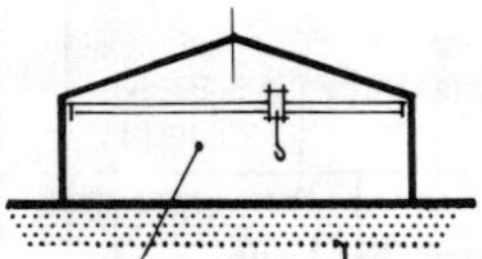

ROOM   ⎫
HANGAR ⎬ HALLE, SAAL

E.G. CRANE BAY   KRANHALLE
      MACHINE ROOM   MASCHINENHALLE
      SPINNING ROOM   SPINNSAAL

SAWTOOTH ROOF STRUCTURE — SHED DACH KONSTRUKTION

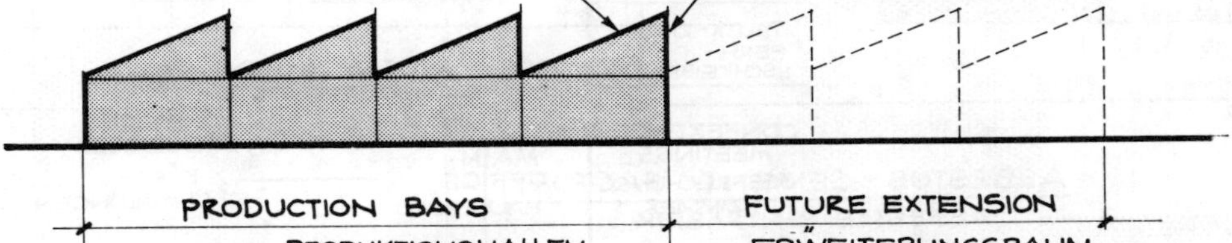

PRODUCTION BAYS   FUTURE EXTENSION
PRODUKTIONSHALLEN   ERWEITERUNGSRAUM

EXISTING — VORHANDEN
PROPOSED — GEPLANT
FUTURE   — ZUKÜNFTIG

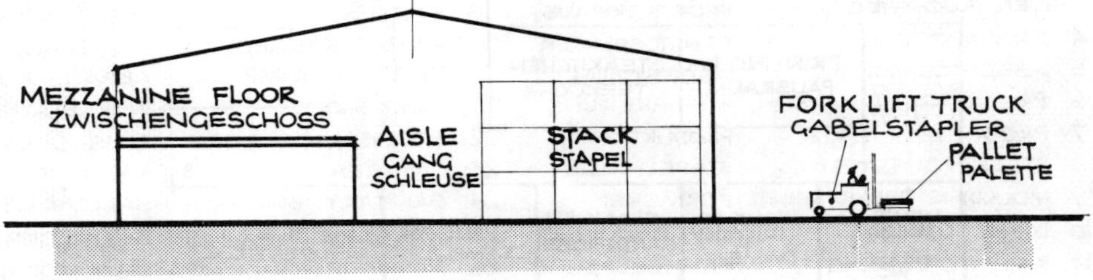

MEZZANINE FLOOR   FORK LIFT TRUCK
ZWISCHENGESCHOSS   AISLE   STACK   GABELSTAPLER
                   GANG   STAPEL   PALLET
                   SCHLEUSE   PALETTE

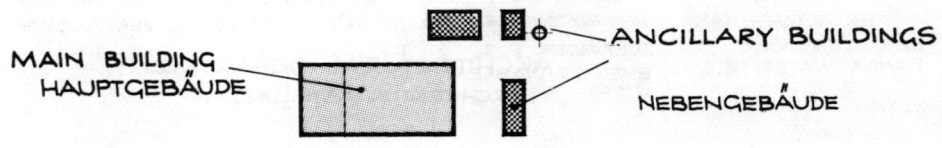

MAIN BUILDING   ANCILLARY BUILDINGS
HAUPTGEBÄUDE   NEBENGEBÄUDE

ARCHITEKTUR / ARCHITECTURE

EXAMPLES:

AN INDUSTRIAL PLANT
EINE INDUSTRIEANLAGE

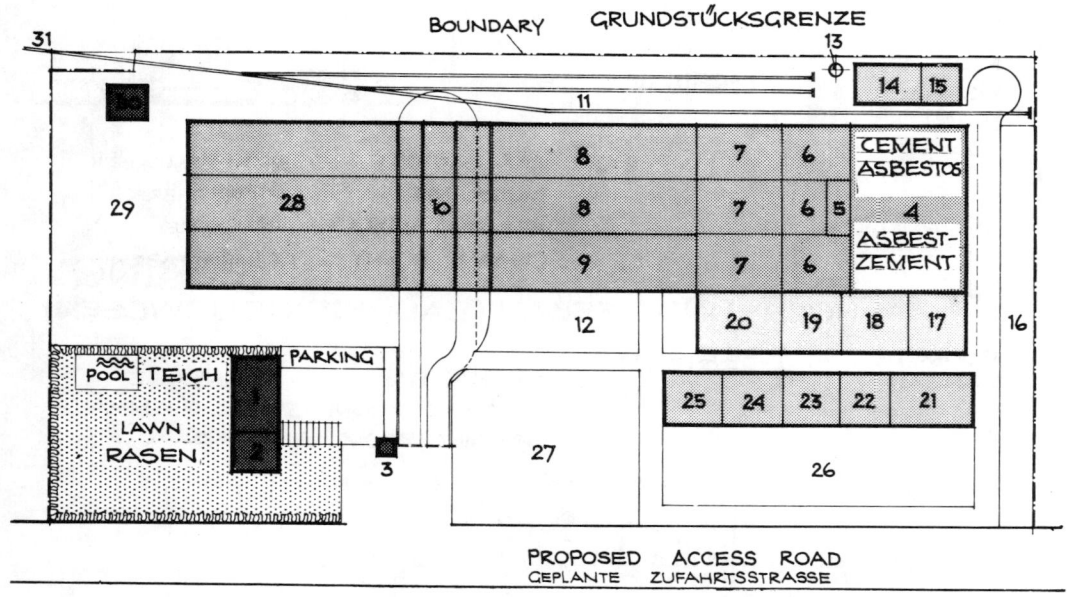

ASBESTOS – CEMENT FACTORY
ASBESTZEMENTFABRIK

LEGEND:

1	ADMINISTRATION	VERWALTUNG
2	HEAD OFFICE	GESCHÄFTSLEITUNG
3	GATE KEEPER'S OFFICE	PFÖRTNERHAUS
4	RAW MATERIAL STORE	ROHSTOFFLAGER
5	ASBESTOS MILLING	ASBESTMÜHLE
6	PREPARATION BAYS	AUFBEREITUNG
7	PRODUCTION BAYS	PRODUKTION
8	STACKING BAYS	STAPELHALLEN
9	MOULDING DEPARTMENT	FORMEREI
10	LORRY LOADING	LKW BELADUNG
11	RAILWAY LOADING	BAHNBELADUNG
12	MOULDS STOCKYARD	FORMENLAGER
13	CHIMNEY	KAMIN
14	BOILER HOUSE	KESSELHAUS
15	WATER STORAGE TANK	WASSERRESERVOIR
16	SERVICES FOR POWER, WATER ETC.	ZUBRINGER FÜR ENERGIE, WASSER U.S.W.
17	SUBSTATION	TRAFOSTATION
18	SETTLING TANKS	ABSETZBECKEN
19	HARDWASTE RECOVERY	ABFALLAUFBEREITUNG
20	RUBBISH DUMP	ABFALLHAUFEN
21	WORKSHOPS	WERKSTÄTTEN
22	ACCESSORIES STORE	ZUBEHÖRLAGER
23	CANTEEN	KANTINE
24	SANITARY BUILDING	SANITÄRGEBÄUDE
25	WELFARE BUILDING	SOZIALGEBÄUDE
26	WORKSHOP STORAGE AREA	WERKSTOFFLAGER
27	MOULDED PRODUCTS	FORMSTÜCKE
28	COVERED STORAGE AREA	ÜBERDACHTES LAGER
29	OPEN STORAGE AREA	OFFENER LAGERPLATZ
30	DISPATCH CONTROL	VERSANDKONTROLLE
31	RAILWAY SIDING	NEBENGLEIS

BUILDING AND CIVIL ENGINEERING
HOCH - UND TIEFBAU (BAUINGENIEURWESEN)

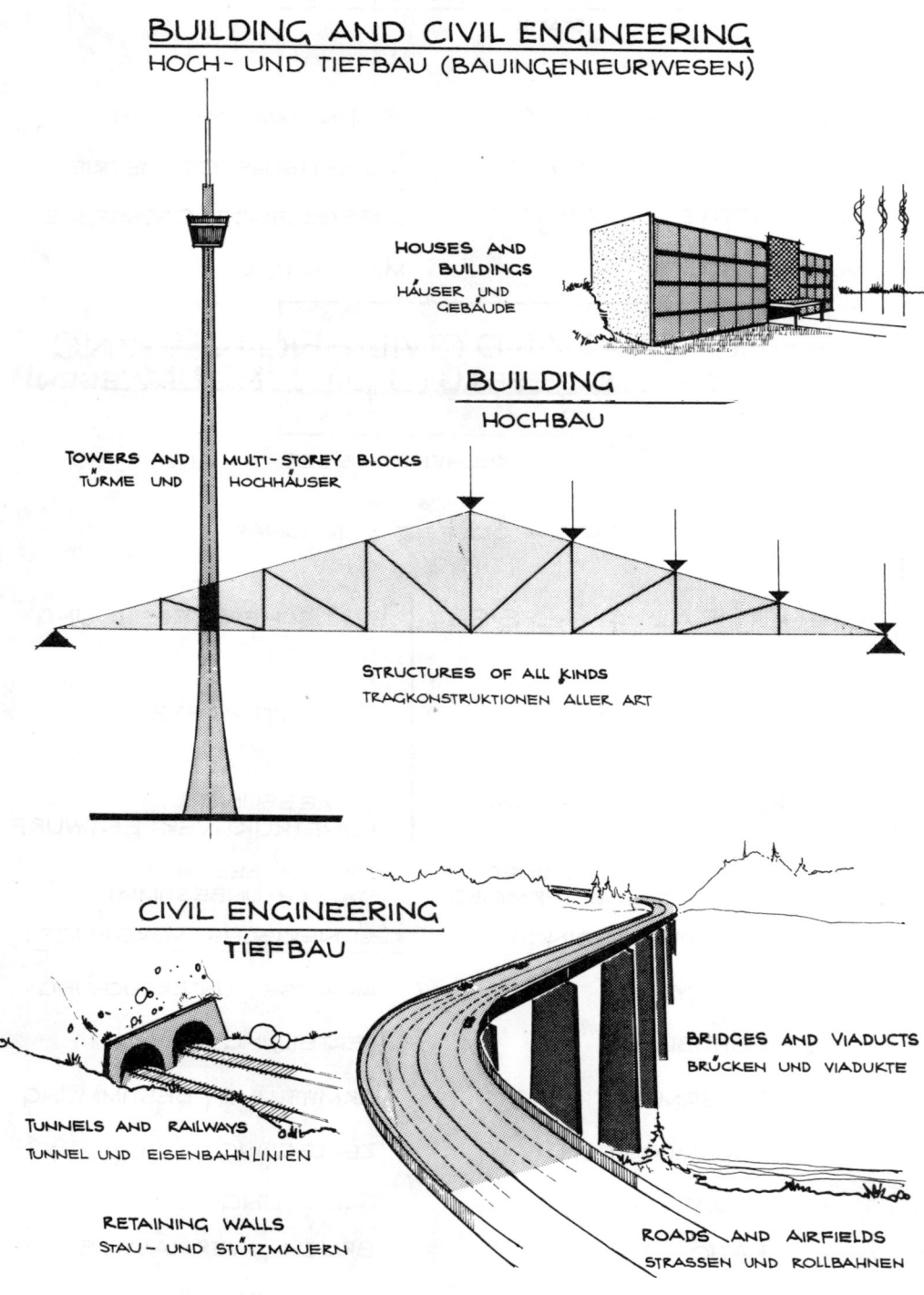

BAUINGENIEURWESEN   CIVIL ENGINEERING

Theory of Structures
Statik

Theory of Structures	Statik (als Lehrfach)
Analytic Geometry	Analytische Geometrie
Descriptive Geometry	Darstellende Geometrie
Mathematics	Mathematik

SLIDE RULE   RECHENSCHIEBER

NOTE: STATIC — ORTSFEST, STATIONÄR

STRUCTURAL ANALYSIS	STATISCHE BERECHNUNG
INTERNAL FORCES	SCHNITTKRÄFTE
STRUCTURAL DESIGN	BEMESSUNG, KONSTRUKTIVER ENTWURF
STATICALLY DETERMINED	STATISCH BESTIMMT
STATICALLY UNDETERMINED	STATISCH UNBESTIMMT
MATHEMATICAL PROCEEDINGS:	RECHNERISCHE VORGÄNGE:
ANALYSIS	ANALYSE, UNTERSUCHUNG
DISTRIBUTION	VERTEILUNG
DETERMINATION	ERMITTLUNG, BESTIMMUNG
DECOMPOSITION	ZERLEGUNG
EQUATION	GLEICHUNG
RATIO	BRUCH, VERHÄLTNIS

BAUINGENIEURWESEN   CIVIL ENGINEERING

GEOMETRY
GEOMETRIE

HORIZONTAL	VERTICAL	INCLINED
HORIZONTAL, WAAGRECHT	VERTIKAL, SENKRECHT	SCHIEF

SPACE	—	RAUM		CHORD	—	SEHNE
POINT	—	PUNKT		TANGENT	—	TANGENTE
TERM	—	GLIED		PLANE	—	EBENE
LINE	—	GERADE		AREA	—	FLÄCHE
LOCALITY	—	LAGE		CONTENT, VOLUME	—	RAUMINHALT, VOLUMEN

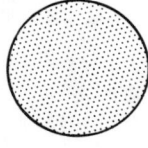 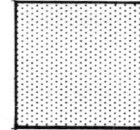 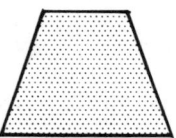 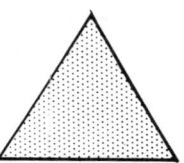

CIRCLE — KREIS **SQUARE** — QUADRAT **TRAPEZIUM** — TRAPEZ **TRIANGLE** — DREIECK

RECTANGLE — RECHTECK

SYMMETRICALLY ANGLED	—	GLEICHWINKLIG
RIGHT ANGLED	—	RECHTWINKLIG
OBLIQUE ANGLED	—	SCHIEFWINKLIG
ACUTE ANGLED	—	SPITZWINKLIG
OBTUSE ANGLED	—	STUMPFWINKLIG
ISOSCELES	—	GLEICHSCHENKLIG
ASOSCELES	—	UNGLEICHSCHENKLIG
WITH EQUAL SIDES	—	GLEICHSEITIG
SCALENE / WITH UNEQUAL SIDES	—	UNGLEICHSEITIG

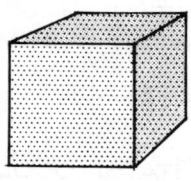 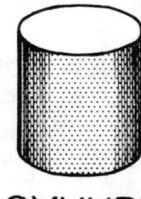 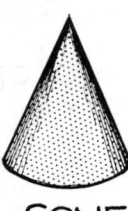 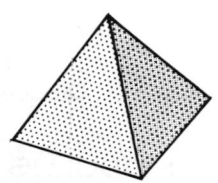

CUBE — WÜRFEL **CYLINDER** — ZYLINDER **CONE** — KEGEL **PYRAMID** — PYRAMIDE

BAUINGENIEURWESEN CIVIL ENGINEERING

PROPERTIES OF MATERIAL
MATERIALEIGENSCHAFTEN

SIZE GRÖSSE

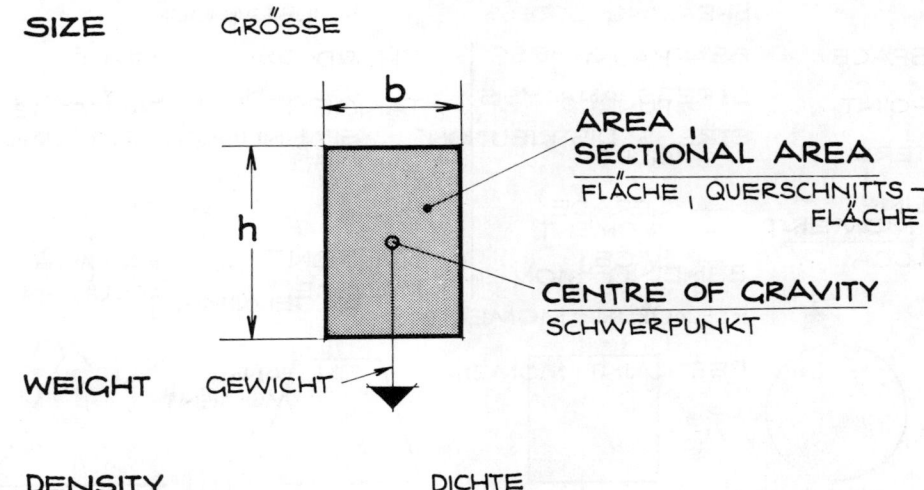

AREA, SECTIONAL AREA
FLÄCHE, QUERSCHNITTS-FLÄCHE

CENTRE OF GRAVITY
SCHWERPUNKT

WEIGHT GEWICHT

DENSITY DICHTE

STRENGTH, SOLIDITY FESTIGKEIT

FLEXIBILITY BIEGSAMKEIT
CONSISTENCY KONSISTENZ

ELASTIC MODULUS } WIDERSTANDS- $\dfrac{bh^2}{6}$
SECTION MODULUS   MOMENT

MOMENT OF INERTIA } TRÄGHEITS- $\dfrac{bh^3}{12}$
MOMENT OF GYRATION   MOMENT

RADIUS OF GYRATION
TRÄGHEITSRADIUS

MODULUS OF ELASTICITY ELASTIZITÄTSMODUL

BAUINGENIEURWESEN   CIVIL ENGINEERING

STRUCTURAL TERMS
STATISCHE GRÖSSEN

STRESS — SPANNUNG

    SHEARING STRESS   SCHUBSPANNUNG
    BENDING STRESS   BIEGESPANNUNG
    STRESS ANALYSIS   SPANNUNGSERMITTLUNG
    STRESS DISTRIBUTION   SPANNUNGSVERTEILUNG

MOMENT — MOMENT

    BENDING MOMENT   ⎫
    FLEXURAL MOMENT   ⎬ BIEGEMOMENT
    RESTRAINT MOMENT   EINSPANN-MOMENT

SHEAR FORCE — QUERKRAFT

    SHEAR LOAD   SCHERLAST
    SHEAR ACTION   ABSCHEREN
    SHEARING RESISTANCE   SCHUBWIDERSTAND

BEARING PRESSURE — AUFLAGERDRUCK, AUFLAGERPRESSUNG

    SOIL PRESSURE   BODENPRESSUNG

STRENGTH — FESTIGKEIT

    FLEXURAL STRENGTH   BIEGEZUGFESTIGKEIT
    SHEARING STRENGTH   SCHUBFESTIGKEIT

BAUINGENIEURWESEN   CIVIL   ENGINEERING

TYPES OF LOADS
LASTARTEN

NOTE: 1 TON CAN BE A LONG TON (1016,05 kg)
                   OR A SHORT TON (907,18 kg)
                   OR A METRIC TON (1 000,00 kg)

DESIGN LOAD σ LASTANNAHME
PERMISSIBLE LOAD   ZULÄSSIGE BELASTUNG , $\sigma_{zul.}$
ACTUAL LOAD   TATSÄCHLICHE BELASTUNG

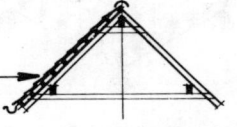

DEAD LOAD
EIGENGEWICHT

LIVE LOADS   :
VERKEHRSLASTEN   :

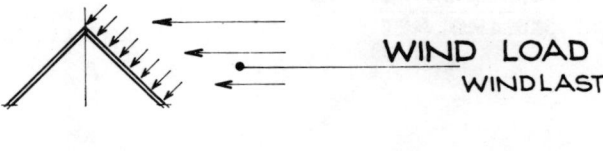

IMPOSED LOAD
NUTZLAST

WIND LOAD
WINDLAST

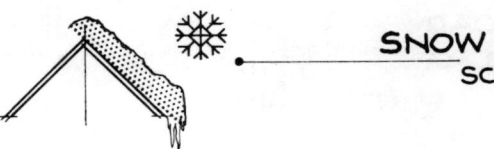

SNOW LOAD
SCHNEELAST

REPAIR LOAD
REPARATURLAST

WHEEL PRESSURE
RADLAST

BAUINGENIEURWESEN — CIVIL ENGINEERING

IMPORTANT IMPOSED LOADS
WICHTIGE NUTZLASTEN, BELASTUNGSFÄLLE

POINT LOAD, SINGLE LOAD, CONCENTRATED LOAD
EINZELLAST, PUNKTLAST
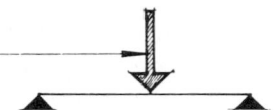

UNIFORMLY DISTRIBUTED LOAD
GLEICHLAST, GLEICHMÄSSIG VERTEILTE LAST

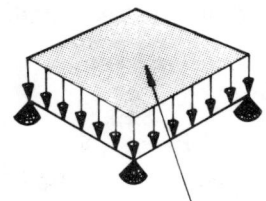
UNIFORMLY DISTRIBUTED LOAD ON SLAB
AREA LOAD
FLÄCHENLAST

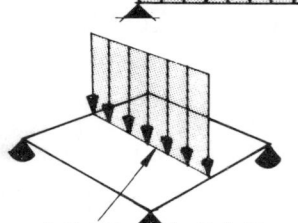
LINE LOAD
LINEAR LOAD
KNIFE - EDGE LOAD
LINIENLAST, STRECKENLAST

TRIANGULAR LOAD
DREIECKSLAST
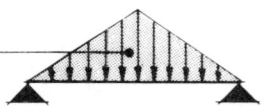

BLOCK LOAD
BLOCKLAST
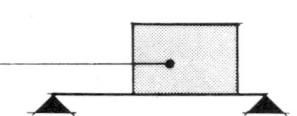

SWELLING LOAD, LOAD INCREASING WITH A CERTAIN FREQUENCY
SCHWELLAST

E.G. PEOPLE IN EQUAL STEP
Z.B. MARSCHIERENDE KOLONNE

SUPERIMPOSED LOAD
AUFLAST
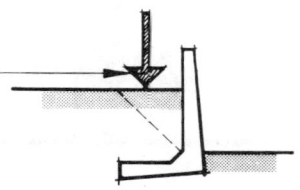

BAUINGENIEURWESEN   CIVIL ENGINEERING

FORCES
KRÄFTE

LOAD, WEIGHT	GEWICHT, LAST
COUNTERWEIGHT	GEGENGEWICHT
EQUILIBRIUM	GLEICHGEWICHT

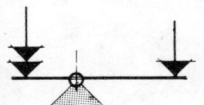

PRESSURE   TENSION
DRUCK   ZUG

SUCTION	SOG
FRICTION	REIBUNG

IMPACT
SCHLAG, ANPRALL

τ
TORSION   TORSION
BOND   HAFTUNG
SHEAR   SCHUB, VERTIKALSCHUBKOMPO-
            NENTE
THRUST   HORIZONTALSCHUBKOMPONENTE

COMPRESSION   SCHEITELDRUCK
LONGITUDINAL COMPRESSION   LÄNGSDRUCK

APPLICATION OF FORCE
KRAFTANGRIFF

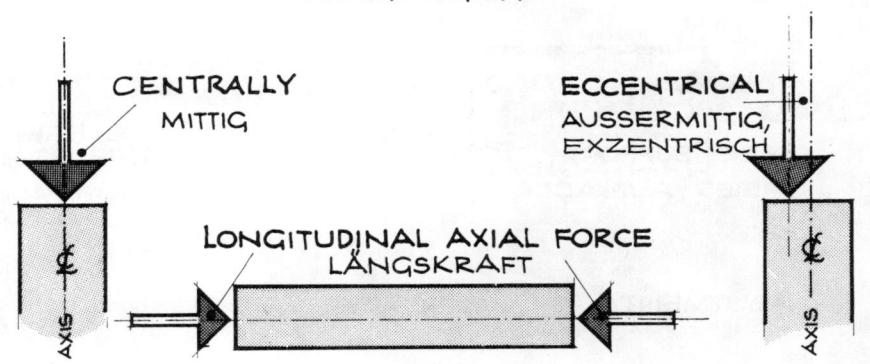

CENTRALLY — MITTIG
ECCENTRICAL — AUSSERMITTIG, EXZENTRISCH
LONGITUDINAL AXIAL FORCE — LÄNGSKRAFT

BAUINGENIEURWESEN — CIVIL ENGINEERING

SPANS
SPANNWEITEN

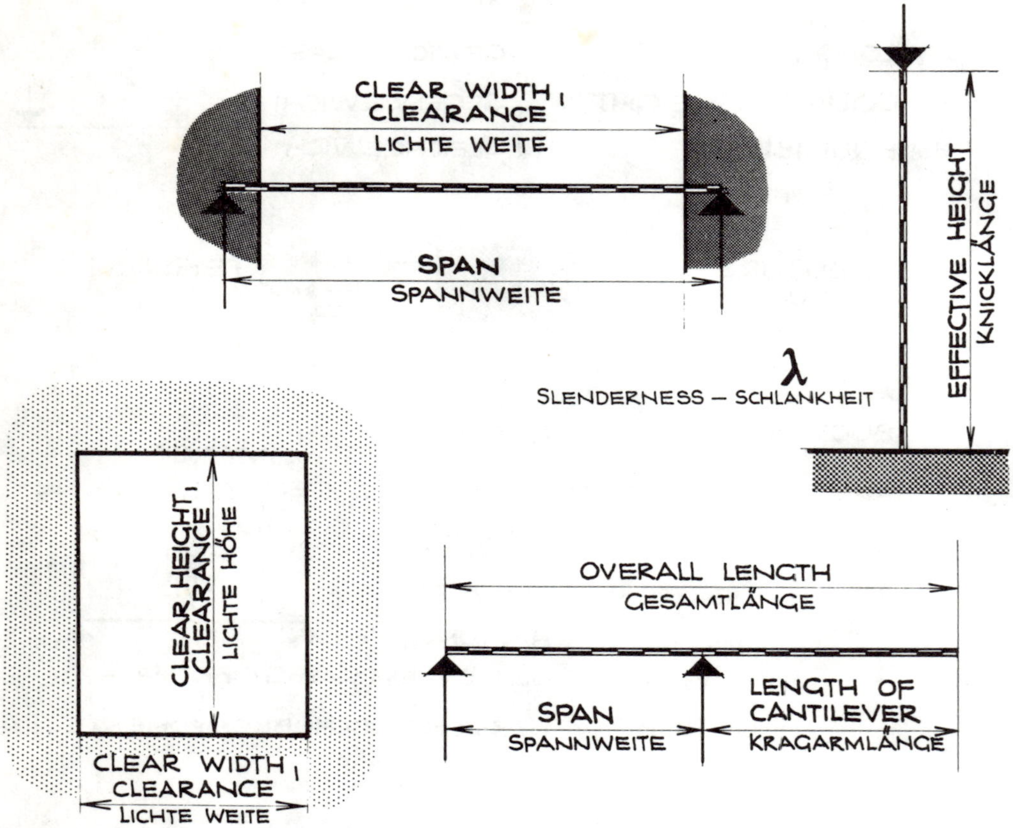

CLEAR WIDTH, CLEARANCE — LICHTE WEITE

SPAN — SPANNWEITE

CLEAR HEIGHT, CLEARANCE — LICHTE HÖHE
CLEAR WIDTH, CLEARANCE — LICHTE WEITE

EFFECTIVE HEIGHT — KNICKLÄNGE

λ SLENDERNESS — SCHLANKHEIT

OVERALL LENGTH — GESAMTLÄNGE
SPAN — SPANNWEITE
LENGTH OF CANTILEVER — KRAGARMLÄNGE

SUPPORTS
AUFLAGER

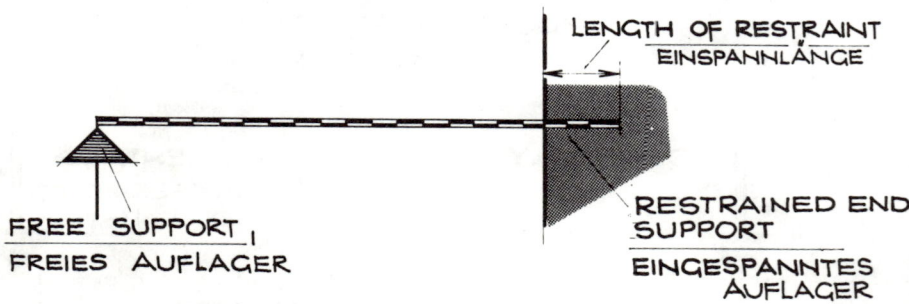

LENGTH OF RESTRAINT — EINSPANNLÄNGE

FREE SUPPORT — FREIES AUFLAGER

RESTRAINED END SUPPORT — EINGESPANNTES AUFLAGER

ABUTMENT — WIDERLAGER

BAUINGENIEURWESEN   CIVIL ENGINEERING

DEFORMATIONS
VERFORMUNGEN

BENDING , FLEXION — BIEGUNG

DEFLECTION — DURCHBIEGUNG

DEFLECTION IN TWO DIRECTIONS
— DURCHBIEGUNG IN ZWEI RICHTUNGEN
  DOPPELBIEGUNG

EXPANSION , DILATATION
— DEHNUNG , AUSDEHNUNG

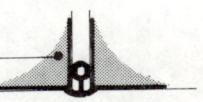

CONTRACTION , SHRINKAGE
— SCHWUND

SWELLING — SCHWELLEN

BUCKLING
LATERAL FLEXURE   } KNICKUNG

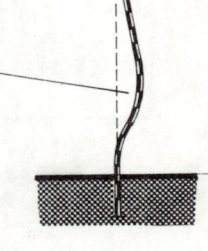

FLEXURAL BUCKLING — BIEGEKNICKUNG

STABILITIES
STANDSICHERHEITEN

STABILITY AGAINST **COLLAPSE**
— STANDSICHERHEIT

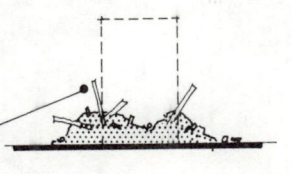

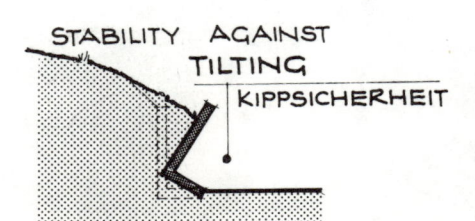

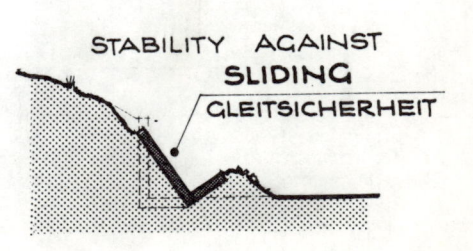

STABILITY AGAINST **TILTING**
— KIPPSICHERHEIT

STABILITY AGAINST **SLIDING**
— GLEITSICHERHEIT

BUILDING KNOWLEDGE
BAUKUNDE

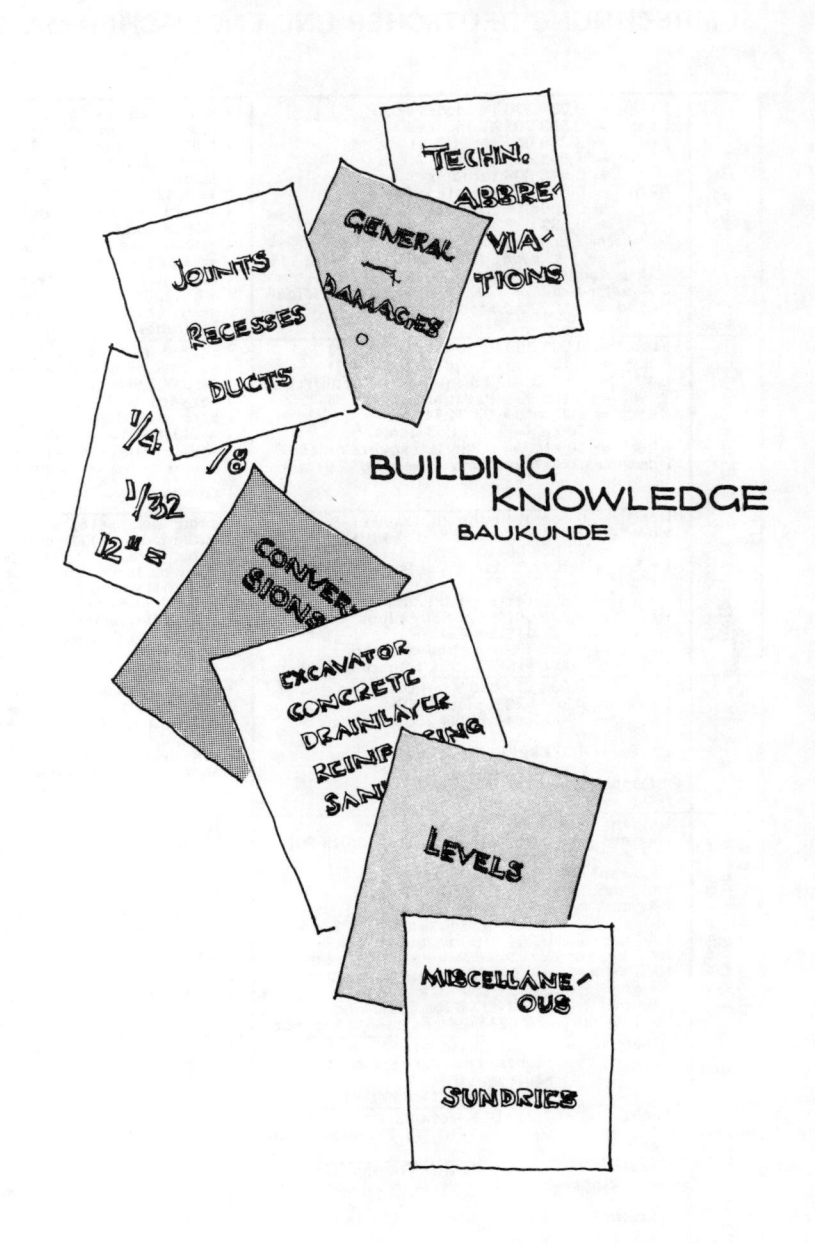

- Techn. Abbreviations
- General Damages
- Joints / Recesses / Ducts
- 1/4 1/8 1/32 12" =
- Conversions
- Excavator / Concrete / Drainlayer / Reinforcing / Sand
- Levels
- Miscellaneous
- Sundries

UMRECHNUNG DEUTSCHER UND ENGLISCHER MASSEINHEITEN

Längenmaße	1 mm = 0,039 370 14 inches (Zoll) 1 cm = 0,393 701 47 inches 1 m = 3,280 851 feet (Fuß) = 1,093 616 yards = 0,546 808 fathoms 1 km = 0,621 372 statute miles = 0,539 614 nautical miles = 0,539 037 admiralty miles *1 deutsche Landmeile = 7,5 km; 1 deutsche Seemeile = 1,852 km; 1 geographische Meile (15 = 1 Äquatorgrad) = 7,420 438 54 km 1 Äquator-Grad = 111,306 6 km; 1 Meridian-Grad = 111,120 6 km*	1 inch = **25,399 956 mm** ... [1]) *1 line = 0,1 inches* 1 foot = 12 in. = 304,799 472 mm = 0,304 799 m 1 yard = 3 feet = 36 in. = 0,914 398 m 1 fathom = 2 yards = 6 feet = 72 in. = 1,828 797 m 1 stat. mile = 880 fath. = 1760 yards = 5280 feet = 1 engl. Meile = 1,609 341 km 1 gewöhnl. engl. Meile = 5000 feet = 1,523 997 km 1 naut. mile = 6080 feet = 1,853 181 km 1 adm. mile = 6086,5 feet = 1,855 162 km = 1/4 geographische Meile = 1/60 des Äquatorgrades
Flächenmaße	1 mm² = 0,001 550 01 square in. (Zoll²) 1 cm² = 0,155 006 35 square in. (Zoll²) 1 m² = 10,763 983 28 square feet (Fuß²) = 1,195 995 96 square yards 1 a = 100 m² = 0,024 711 acres 1 ha = 100 a = 2,471 063 acres 1 km² = 100 ha = 0,386 100 square miles *1 geographische Quadratmeile = 55,062 91 km²*	1 square inch. = 6,451 578 cm² 1 sq. foot = 144 sq. in. = 929,027 2 cm² = 0,092 903 m² 1 sq. yard = 9 sq. feet = 8 361,244 80 cm² 1 acre = 160 square poles = 4840 square yards = 40,468 4 a = 4 046,842 5 m² 1 square mile = 640 acres = 2,589 979 km² 1 square poles = 25,298 676 m² 1 circular in. = $\frac{\pi}{4}$ sq. in. = 5,067 07 cm²
Körpermaße	1 cm³ = 0,061 024 cubic in. (Zoll³) 1 dm³ = 0,035 315 cubic feet (Fuß³) = 61,024 061 cb. in. 1 m³ = 1,307 957 cubic yards = 35,314 850 feet³ = 0,353 148 Register tons 1 l = 0,220 097 Imperial gallons = 0,027 512 Bushels = 0,003 439 Imperial quarters 1 hl = 100 l = 0,343 901 Imperial quarters	1 cubic inch = 16,386 979 cm³ 1 cubic foot = 1728 cubic in. = 28,316 700 dm³ 1 cubic yard = 27 cubic feet = 0,764 551 m³ 1 Register ton = 100 cubic feet = 2,831 670 m³ 1 Imperial gallon = 277,26 cub. in. = 4,543 454 l 1 Bushel = 8 Imp. gall. = 36,366 7 l 1 Imperial quarter = 8 Bushel = 64 Imp. gall. = 290,941 63 l = 2,909 42 hl
Gewichte	1 g = 0,035 273 956 8 ounces 1 kg = 2,204 622 3 pounds (lbs) 1 t = 1000 kg = 0,984 206 long tons 1 t = 1000 kg = 1,102 311 short tons (Schiffstonne) *1 Doppelzentner (dz) = 100 kg* *100 g = 1 Hektogramm*	1 ounce = 28,349 53 g 1 pound (lb) = 16 ounces = 0,453 592 4 kg 1 long ton = 20 centweights = 2240 lbs = 1 016,047 06 kg = 1,12 short tons 1 centweight (cwt) = 112 lbs = 50,802 349 kg 1 short ton = 2000 lbs = 907,184 88 kg = 0,892 857 long ton
Gewichte, bezogen auf Längen-, Flächen- u. Körpermaße	1 kg/cm = 5,599 731 lbs/in. 1 kg/m = 0,671 967 lbs/foot = 2,015 901 lbs/yard 1 kg/mm² = 0,634 968 long tons/sq. in. = 1 422,329 261 lbs/sq. in. 1 kg/cm² = 0,006 349 6 long tons/sq. in. = 14,223 335 lbs/in.² 1 kg/m² = 0,204 816 lbs/square foot 1000 kg/cm² = 6,349 684 long tons/square in. 1000 kg/m² = 0,091 435 5 long tons/square foot 1 kg/cm³ = 36,127 168 lbs/cubic in. 1 kg/m³ = 0,062 427 6 lbs/cubic foot 1000 kg/m³ = 0,027 869 long tons/cubic foot	1 lbs/inch = 0,178 580 kg/cm 1 lbs/foot = 1,488 169 kg/m 1 lbs/yard = 0,496 056 kg/m 1 lbs/square in. = 0,070 307 kg/cm² 1 lbs/square foot = 4,882 430 kg/m² 1 long ton/square in. = 157,488 146 kg/cm² 1 long ton/square foot = 10,936 677 t/m² 1 lbs/cubic in. = 0,027 680 kg/cm³ 1 lbs/cubic foot = 16,018 549 kg/m³ 1 long ton/cubic foot = 35,881 549 t/m³
Druck	1 metr. at = 14,223 lbs/square in. = 28,958 inches of mercury (Quecksilbersäule von 0°) = 393,7 inches of water (Wassersäule)	1 pound/square inch = 0,070 307 metr. at
Arbeit	1 mkg = 7,233 011 feet pounds = 0,009 295 British Thermal Units (BTU)	1 foot pound = 0,138 255 mkg 1 BTU = 107,58 mkg
Wärme	1 kcal = 3,968 3 British Thermal Units 1 kcal/cm² = 25,60 British Thermal Units/square in. 1 kcal/m³ = 0,112 368 British Thermal Units/cubic foot $\frac{9 C°}{5} + 32$ = F° (Fahrenheit)	1 BTU = 0,252 kcal (W.-E.) 1 BTU/sq. inch = 0,039 060 kcal/cm² 1 BTU/cubic foot = 8,899 342 kcal/m³ $\frac{5 (F° - 32)}{9}$ = C° (Celsius)
	1 cm⁴ = 0,024 025 26 inches⁴ 1 m/s = 196,851 feet/min	1 inch⁴ = 41,622 854 cm⁴ 1 foot/min = 0,005 08 m/s

[1]) Nach DIN 4890 Bl 1 gesetzlich festgelegter Vergleichswert. **Allgemein gilt für Werkstücke und Meßgeräte für den englischen und für den amerikanischen Zoll**
1″ = 25, 400 000 mm.
Nur für Maße und Messungen allerhöchster Genauigkeit ist der gesetzlich festgelegte Wert zugrunde zu legen und zu rechnen:
1″ engl. = 25,399 956 mm
1″ amerik. = 25,400 051 mm.

	Birmingham Wire			United States Standard				
	Thickness				Approximate Thickness, Inch			Weight per Square Foot Pounds, Av.
Number	Decimal Inches	Nearest 1/64 Inch	Millimeters	Number	Steel	Iron	Nearest 1/64 Inch	
0000	0.454	29/64	11.532	1	0.2757	0.2813	9/32	11.25
000	0.425	27/64	10.795	2	0.2604	0.2656	17/64	10.625
00	0.380	3/8	9.652	3	0.2451	0.25	1/4	10.00
0	0.340	11/32	8.636	4	0.2298	0.2344	15/64	9.375
				5	0.2145	0.2188	7/32	8.75
1	0.300	19/64	7.620	6	0.1991	0.2031	13/64	8.125
2	0.284	9/32	7.214	7	0.1838	0.1875	3/16	7.5
3	0.259	17/64	6.579	8	0.1685	0.1719	11/64	6.875
4	0.238	15/64	6.045	9	0.1532	0.1563	5/32	6.25
5	0.220	7/32	5.588	10	0.1379	0.1406	9/64	5.625
6	0.203	13/64	5.156	11	0.1225	0.125	1/8	5.00
7	0.180	3/16	4.572	12	0.1072	0.1094	7/64	4.375
8	0.165	11/64	4.191	13	0.0919	0.0938	3/32	3.75
9	0.148	9/64	3.759	14	0.0766	0.0781	5/64	3.125
10	0.134	9/64	3.404	15	0.0689	0.0703	1/16	2.8125
11	0.120	1/8	3.048	16	0.0613	0.0625	1/16	2.50
12	0.109	7/64	2.769	17	0.0551	0.0563	1/16	2.25
13	0.095	3/32	2.413	18	0.0490	0.05	3/64	2.00
14	0.083	5/64	2.108	19	0.0429	0.0438	3/64	1.75
15	0.072	5/64	1.829	20	0.0368	0.0375	1/32	1.50
16	0.065	1/16	1.651	21	0.0337	0.0344	1/32	1.375
17	0.058	1/16	1.473	22	0.0306	0.0313	1/32	1.25
18	0.049	3/64	1.245	23	0.0276	0.0281	1/32	1.125
19	0.042	3/64	1.067	24	0.0245	0.025	1/32	1.00
20	0.035	1/32	0.889	25	0.0214	0.0219	1/64	0.875
21	0.032	1/32	0.813	26	0.0184	0.0188	1/64	0.75
22	0.028	1/32	0.711	27	0.0169	0.0172	1/64	0.6875
23	0.025	1/32	0.635	28	0.0153	0.0156	1/64	0.625
24	0.022	1/64	0.559	29	0.0138	0.0141	....	0.5625
25	0.020	1/64	0.508	30	0.0123	0.0125	....	0.50
26	0.018	1/64	0.457	31	0.0107	0.0109	....	0.4375
27	0.016	1/64	0.406	32	0.0100	0.0102	....	0.40625
28	0.014	....	0.356	33	0.0092	0.0094	....	0.375
29	0.013	....	0.330	34	0.0084	0.0086	....	0.34375
30	0.012	....	0.305	35	0.0077	0.0078	....	0.3125
31	0.010	....	0.254	36	0.0069	0.0070	....	0.28125
32	0.009	....	0.229	37	0.0065	0.0066	....	0.265625
33	0.008	....	0.203	38	0.0061	0.0063	....	0.25
34	0.007	....	0.178	39	0.0057	....	....	0.234375
35	0.005	....	0.127	40	0.0054	....	....	0.21875
36	0.004	....	0.102	41	0.0052	....	....	0.2109375
				42	0.0050	....	....	0.203125
				43	0.0048	....	....	0.1953125
				44	0.0046	....	....	0.1875

BAUKUNDE   BUILDING KNOWLEDGE

CHART
ORGANISATIONSSCHEMA

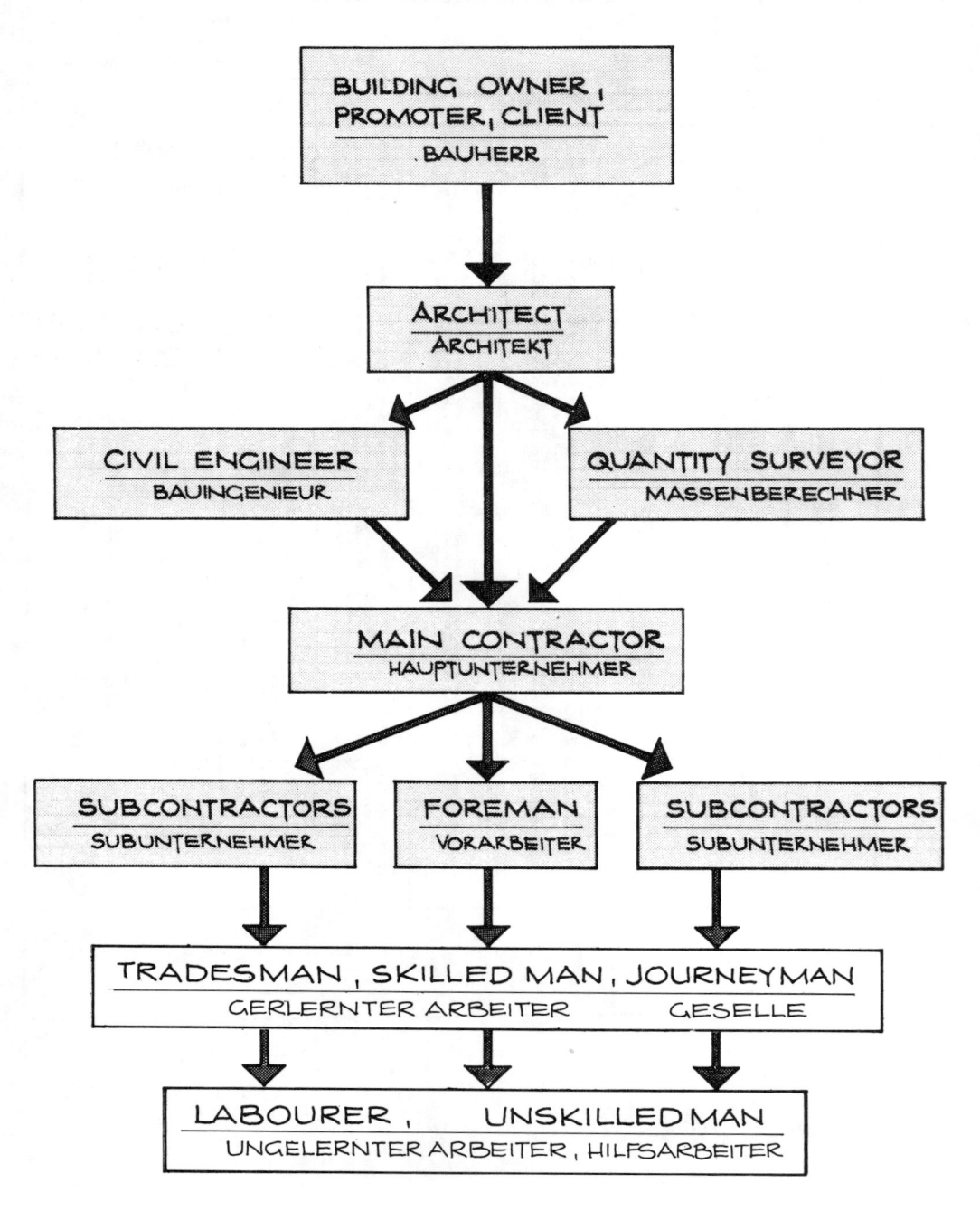

```
┌─────────────────────────────┐
│ BUILDING OWNER,             │
│ PROMOTER, CLIENT            │
│ BAUHERR                     │
└─────────────────────────────┘
              │
              ▼
      ┌───────────────┐
      │ ARCHITECT     │
      │ ARCHITEKT     │
      └───────────────┘
        │           │
        ▼           ▼
┌──────────────┐  ┌────────────────────┐
│ CIVIL ENGINEER│  │ QUANTITY SURVEYOR  │
│ BAUINGENIEUR  │  │ MASSENBERECHNER    │
└──────────────┘  └────────────────────┘
        │           │
        ▼           ▼
      ┌──────────────────┐
      │ MAIN CONTRACTOR  │
      │ HAUPTUNTERNEHMER │
      └──────────────────┘
       │       │       │
       ▼       ▼       ▼
┌───────────┐ ┌────────┐ ┌───────────┐
│SUBCONTRAC-│ │FOREMAN │ │SUBCONTRAC-│
│TORS       │ │VORAR-  │ │TORS       │
│SUBUNTER-  │ │BEITER  │ │SUBUNTER-  │
│NEHMER     │ │        │ │NEHMER     │
└───────────┘ └────────┘ └───────────┘
       │        │          │
       ▼        ▼          ▼
┌──────────────────────────────────────┐
│ TRADESMAN, SKILLED MAN, JOURNEYMAN   │
│ GERLERNTER ARBEITER     GESELLE      │
└──────────────────────────────────────┘
       │        │          │
       ▼        ▼          ▼
┌──────────────────────────────────────┐
│ LABOURER,    UNSKILLED MAN           │
│ UNGELERNTER ARBEITER, HILFSARBEITER  │
└──────────────────────────────────────┘
```

BAUKUNDE   BUILDING KNOWLEDGE

BUILDING CRAFTSMEN } BAUHANDWERKER
BUILDING TRADESMAN

BUILDING TRADE :   BAUGEWERBE :

BRICKLAYER   ZIEGELMAURER
MASON   MAURER

CONCRETE WORKER   BETONBAUER

STEEL BENDER   BETONSTAHLBIEGER
STEEL FIXER   BETONSTAHLVERLEGER

PLASTERER   VERPUTZER

CARPENTER   ZIMMERMANN

SHEETMETAL WORKER,   SPENGLER
TIN SMITH

PLUMBER   INSTALLATEUR

ELECTRICIAN   ELEKTRIKER

SMITH, LOCKSMITH   SCHLOSSER

JOINER   SCHREINER
IRONMONGER   BESCHLÄGESPEZIALIST
GLAZIER   GLASER

TILER   FLIESENLEGER

FLOOR LAYER   BODENLEGER

PAINTER   MALER

PAPERHANGER   TAPEZIERER

35

BAUKUNDE BUILDING KNOWLEDGE

BUILDING WORK
BAUARBEITEN

EXCAVATION WORK	AUSHUBARBEIT
CONCRETE WORK	BETONARBEIT
FORMWORK / SHUTTERING WORK	SCHALARBEIT
REINFORCING WORK	BEWEHRUNGSARBEIT
CARPENTRY WORK	ZIMMERMANNSARBEIT
SHEETMETAL WORK	SPENGLERARBEIT
ROOF TILING WORK	DACHDECKERARBEIT
WATER PROOFING WORK	DICHTUNGSARBEIT
PLUMBING WORK	SANITÄRINSTALLATION
INSULATING WORK	ISOLIERARBEIT
ELECTRICAL INSTALLATION	ELEKTROINSTALLATION
BLACKSMITH WORK	SCHLOSSERARBEIT
PLASTER WORK / RENDERING WORK	VERPUTZARBEIT
JOINERY WORK	SCHREINERARBEIT
IRONMONGERY WORK	BESCHLAGARBEIT
GLAZING WORK	GLASERARBEIT
TILING WORK	FLIESENLEGERARBEIT
TERRAZZO WORK	KUNSTSTEINARBEIT
FLOOR LAYING WORK	BODENLEGERARBEIT
PAINTING WORK	MALERARBEIT
WALLPAPERING WORK	TAPEZIERARBEIT

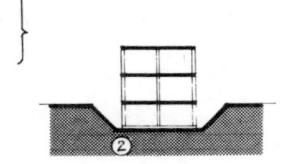

BAUKUNDE   BUILDING KNOWLEDGE

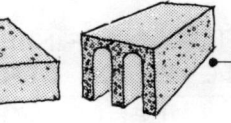

BUILDING MATERIALS
BAUMATERIALIEN

LIME	— KALK
CEMENT	— ZEMENT
GYPSUM, PLASTER OF PARIS	— GIPS

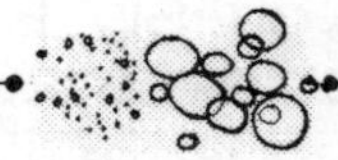

SAND — GRAVEL, STONE
SAND — KIES

BRICK — HOLLOW BLOCK
ZIEGELSTEIN — HOHLBLOCKSTEIN

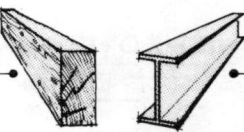

TIMBER — MILD STEEL
BAUHOLZ — FLUSS-STAHL

ROOFING TILE — ROOFING FELT
DACHZIEGEL — DACHPAPPE

REINFORCING STEEL — STEEL PIPE
BEWEHRUNGS-STAHL — STAHLROHR
— PLASTIC PIPE
— PLASTIKROHR

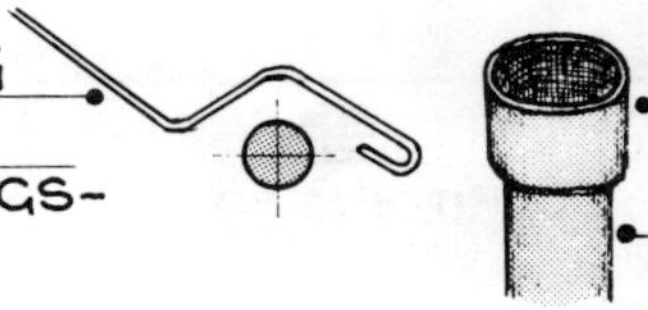

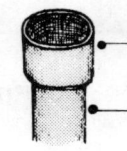

INSULATION MATERIAL	— ISOLIERMATERIAL
PUTTY, MASTIC	— KITT, DICHTUNGSMASSE
TILE, SLATE	— FLIESE
PAINT	— FARBE
WALLPAPER	— TAPETE

BAUKUNDE   BUILDING KNOWLEDGE

LEVELS
HÖHEN, KOTEN

DATUM,
ELEVATION   —   ABSOLUTKOTE
                        KOTE ÜBER NORMAL NULL (ÜNN)

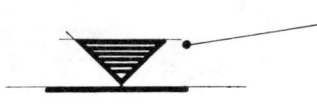

▽   TOP OF SLAB, TOP OF BEAM, ETC.
     LEVEL OF STRUCTURE WITHOUT
     ROHKOTE   FINISH

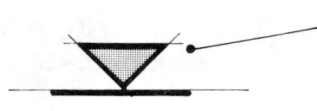

▽   TOP OF FINISH,
     LEVEL OF STRUCTURE WITH FINISH
     FERTIGKOTE

FFL FINISHED FLOOR LEVEL
▽   FUSSBODENOBERKANTE

EXAMPLE:

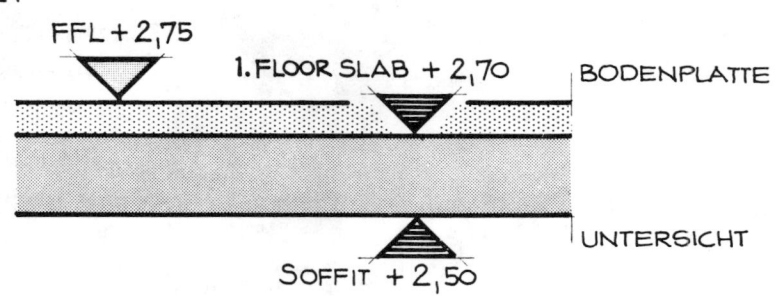

FFL + 2,75
1. FLOOR SLAB + 2,70   BODENPLATTE

SOFFIT + 2,50   UNTERSICHT

TOP LEVEL
OBERKANTE

BOTTOM LEVEL
UNTERKANTE
SOFFIT LEVEL
UNTERSICHT KOTE

BAUKUNDE   BUILDING KNOWLEDGE

JOINTS
VERBINDUNGEN, STÖSSE, FUGEN

JOINTED — VERBUNDEN, GESTOSSEN

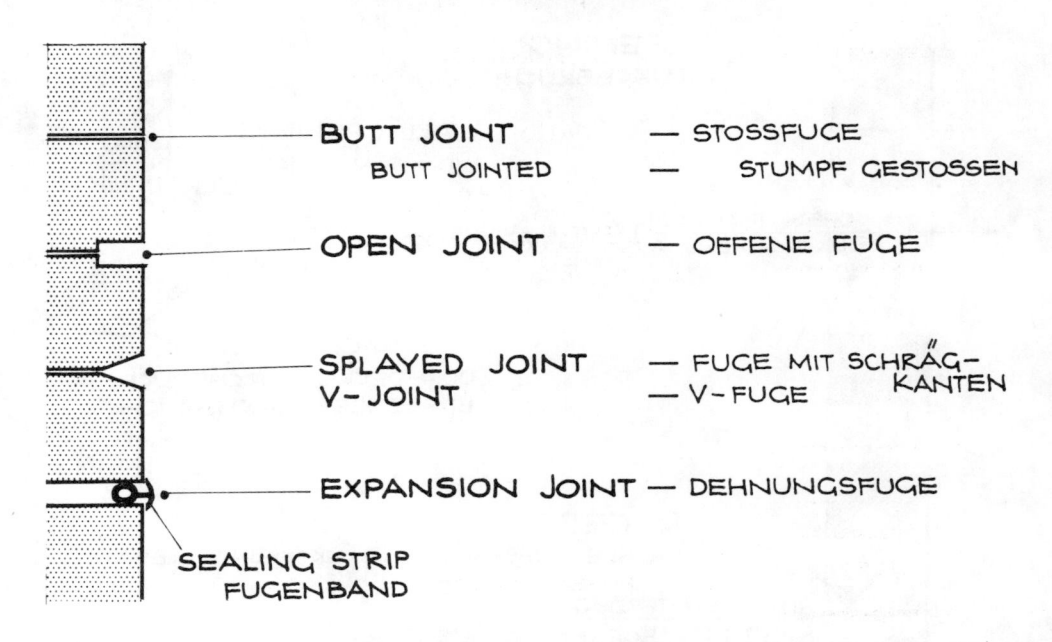

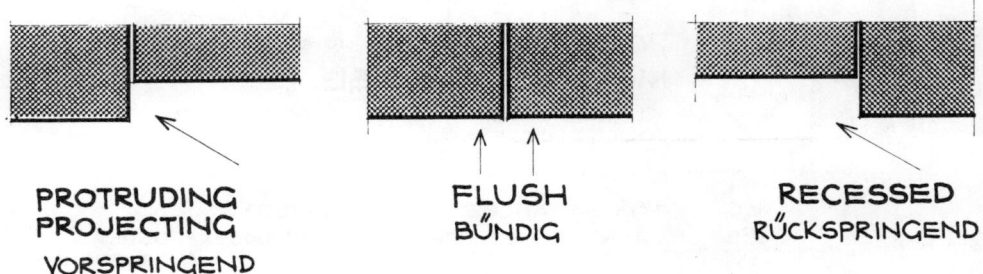

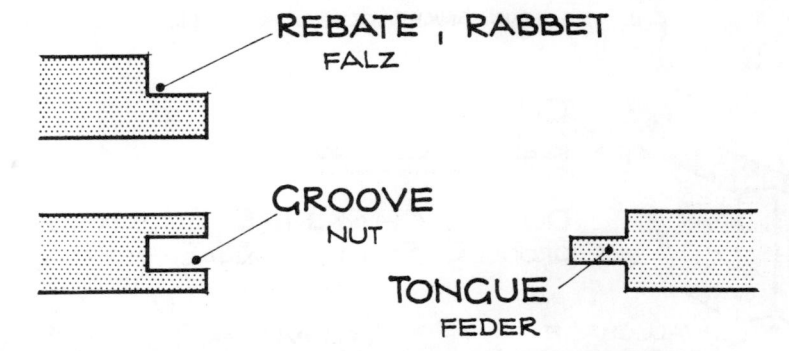

BUTT JOINT — STOSSFUGE
  BUTT JOINTED — STUMPF GESTOSSEN

OPEN JOINT — OFFENE FUGE

SPLAYED JOINT — FUGE MIT SCHRÄG-KANTEN
V-JOINT — V-FUGE

EXPANSION JOINT — DEHNUNGSFUGE

SEALING STRIP
FUGENBAND

PROTRUDING / PROJECTING
VORSPRINGEND

FLUSH
BÜNDIG

RECESSED
RÜCKSPRINGEND

REBATE, RABBET
FALZ

GROOVE
NUT

TONGUE
FEDER

TONGUED AND GROOVED
GENUTET UND GEFEDERT
(MIT NUT UND FEDER)

BAUKUNDE — BUILDING KNOWLEDGE

RECESSES AND OPENINGS
AUSSPARUNGEN

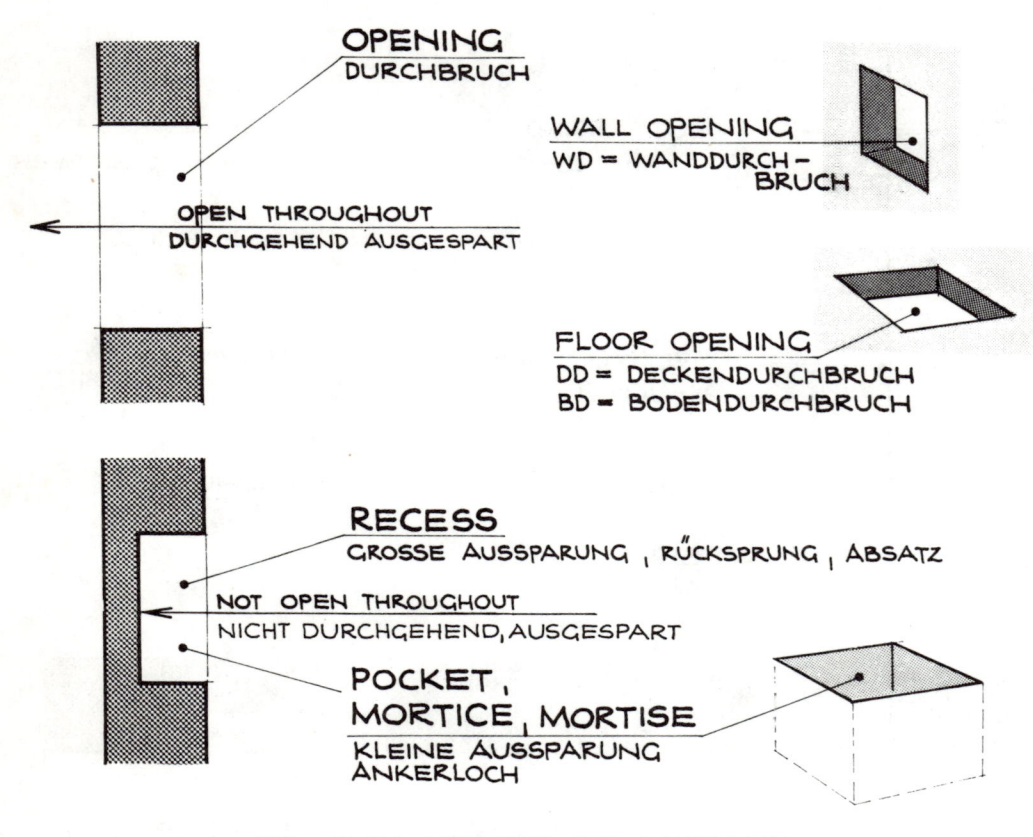

OPENING
DURCHBRUCH

WALL OPENING
WD = WANDDURCH-BRUCH

OPEN THROUGHOUT
DURCHGEHEND AUSGESPART

FLOOR OPENING
DD = DECKENDURCHBRUCH
BD = BODENDURCHBRUCH

RECESS
GROSSE AUSSPARUNG, RÜCKSPRUNG, ABSATZ

NOT OPEN THROUGHOUT
NICHT DURCHGEHEND, AUSGESPART

POCKET, MORTICE, MORTISE
KLEINE AUSSPARUNG
ANKERLOCH

E.G. FORM MORTICES FOR BALUSTERS
Z.B. ES SIND ANKERLÖCHER FÜR GELÄNDERPFOSTEN AUSZUSPAREN

E.G. FORM POCKETS FOR BOLTS
Z.B. ES SIND ANKERLÖCHER FÜR BOLZEN ZU SCHALEN

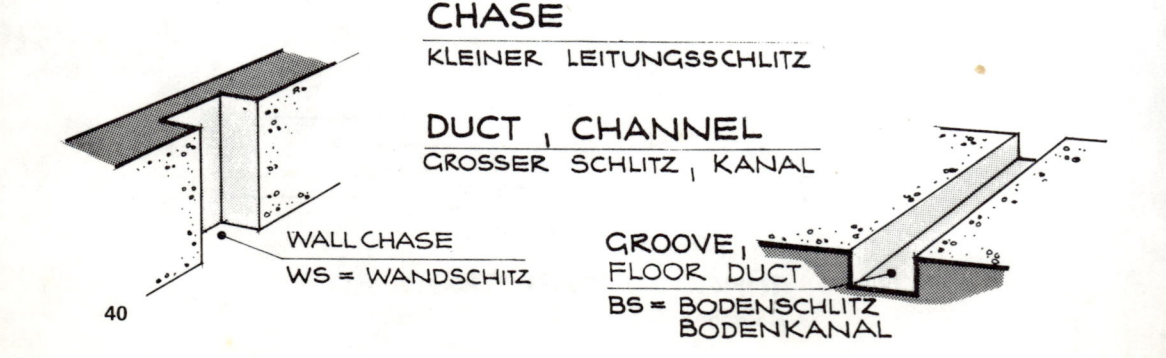

CHASE
KLEINER LEITUNGSSCHLITZ

DUCT, CHANNEL
GROSSER SCHLITZ, KANAL

WALL CHASE
WS = WANDSCHLITZ

GROOVE, FLOOR DUCT
BS = BODENSCHLITZ
BODENKANAL

BAUKUNDE   BUILDING KNOWLEDGE

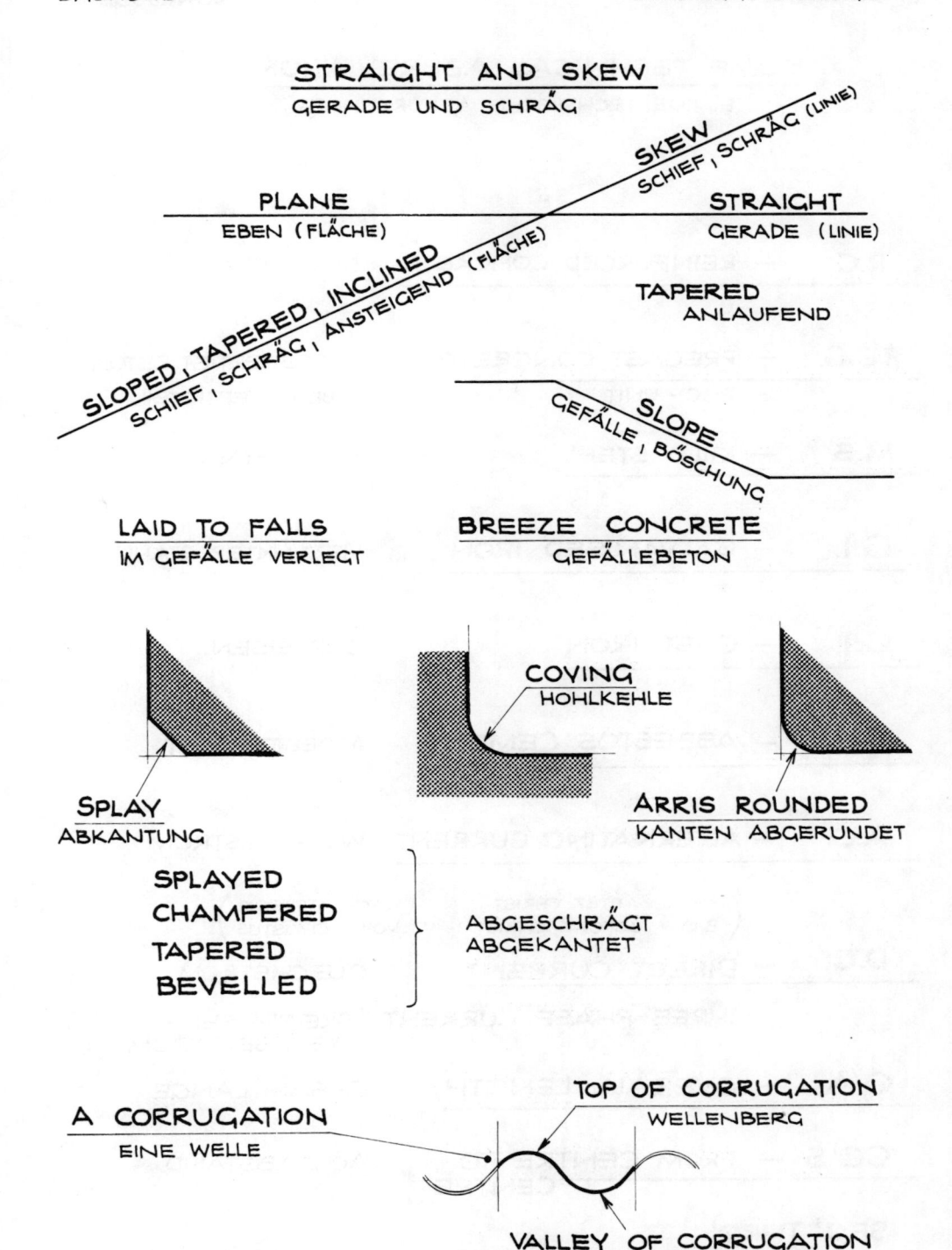

STRAIGHT AND SKEW
GERADE UND SCHRÄG

SKEW
SCHIEF, SCHRÄG (LINIE)

PLANE
EBEN (FLÄCHE)

STRAIGHT
GERADE (LINIE)

SLOPED, TAPERED, INCLINED
SCHIEF, SCHRÄG, ANSTEIGEND (FLÄCHE)

TAPERED
ANLAUFEND

SLOPE
GEFÄLLE, BÖSCHUNG

LAID TO FALLS
IM GEFÄLLE VERLEGT

BREEZE CONCRETE
GEFÄLLEBETON

COVING
HOHLKEHLE

SPLAY
ABKANTUNG

ARRIS ROUNDED
KANTEN ABGERUNDET

SPLAYED
CHAMFERED
TAPERED
BEVELLED

ABGESCHRÄGT
ABGEKANTET

A CORRUGATION
EINE WELLE

TOP OF CORRUGATION
WELLENBERG

VALLEY OF CORRUGATION
WELLENTAL

BAUKUNDE   BUILDING KNOWLEDGE

SOME TECHNICAL ABBREVIATIONS
EINIGE TECHNISCHE ABKÜRZUNGEN

R.C.	— REINFORCED CONCRETE	STAHLBETON
P.C.C.	— PRECAST CONCRETE	VORGEFERTIGT. BETON
	P.C.C. - UNIT	BETONFERTIGTEIL
M.S.	— MILD STEEL	FLUSS-STAHL
G.I.	— GALVANISED IRON	VERZINKTER STAHL
C.I.	— CAST IRON	GUSSEISEN
A.C.	— ASBESTOS CEMENT	ASBESTZEMENT
A.C.	— ALTERNATING CURRENT	WECHSELSTROM
	(A.C. AFTER CHRIST NACH CHRISTUS)	
	(B.C. BEFORE CHRIST VOR CHRISTUS)	
D.C.	— DIRECT CURRENT	GLEICHSTROM
	THREE PHASE CURRENT	DREIPHASEN-WECHSELSTROM
O/A	— OVERALL LENGTH	GESAMTLÄNGE
CC'S	— FROM CENTRE TO CENTRE	ACHSABSTAND
SEC' THRO'	SECTION THROUGH	SCHNITT DURCH .....

BAUKUNDE   BUILDING KNOWLEDGE

GENERAL
ALLGEMEINES

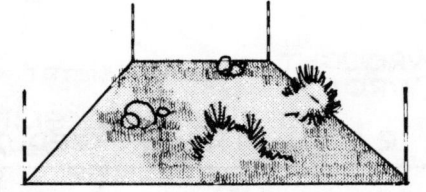

PLOT   BAUGRUNDSTÜCK

ESTATE   BAUGRUNDSTÜCK
                (GROSSEN AUSMASSES)

SITE   BAUSTELLE, GRUNDSTÜCK IN BEBAUUNG

ERECTION   ERRICHTUNG, MONTAGE

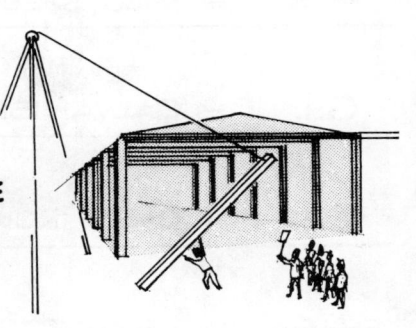

ERECTION OF A BUILDING
ERRICHTUNG EINES GEBÄUDES

INSTALLATION OF A MACHINE
EINBAU EINER MASCHINE

CONSTRUCT A PLANT
EINE ANLAGE ERSTELLEN

TYPES OF DAMAGE   SCHÄDEN

COLLAPSE   EINSTURZ

TILTING   KIPPEN

CRACK   RISS

MOULD, FUNGUS   SCHIMMEL

EFFLORESCENCE   AUSBLÜHUNG

DUST AND SOOT   STAUB UND RUSS

WARP
       VERWERFUNG (HOLZ)

BAUKUNDE   BUILDING KNOWLEDGE

SUNDRIES
SONSTIGES

WROUGHT
WROT   BEARBEITET

AT THE BLACKSMITH'S: HAMMERED, SMITHING (FORGING)
BEIM SCHMIED: GEHÄMMERT, GESCHMIEDET
AT THE JOINER'S: GRINDING, SANDING
BEIM SCHREINER: GESCHLIFFEN, GESCHMIRGELT
AT THE CARPENTER'S: PLANING, DRESSING
BEIM ZIMMERMANN: GEHOBELT, ABGERICHTET

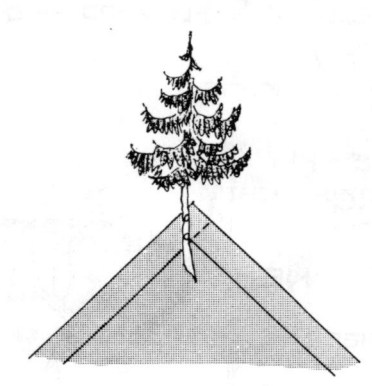

TOPPING-OUT CEREMONY

ROOF WETTING PARTY

RICHTFEST

PROOF :   DICHT :

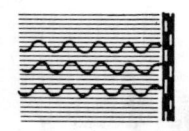

WATER PROOF   WASSERDICHT
MOISTURE PROOF
                     (LUFT)FEUCHTIGKEITSDICHT
DAMP PROOF
                     (KAPILLAR)FEUCHTIGKEITSDICHT

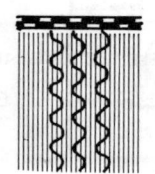

STEAM PROOF   DAMPFDICHT
VAPOUR PROOF   GASDICHT
AIR-TIGHT   LUFTDICHT

VERMIN PROOF   UNGEZIEFERDICHT
SOUND PROOF   SCHALLDICHT

BAUKUNDE   BUILDING KNOWLEDGE

MISCELLANEOUS
VERSCHIEDENES

MARGIN — (TECHNISCH) TOLERANZ, RANDDIFFERENZ, RAND, SPANNE

MARGINAL NOTES — RANDBEMERKUNGEN

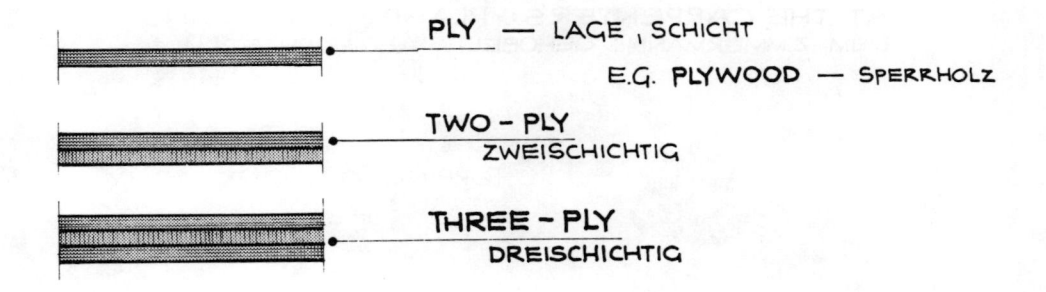

PLY — LAGE, SCHICHT

E.G. PLYWOOD — SPERRHOLZ

TWO-PLY
ZWEISCHICHTIG

THREE-PLY
DREISCHICHTIG

REMOVABLE
DEMOUNTABLE } DEMONTIERBAR, BEWEGLICH

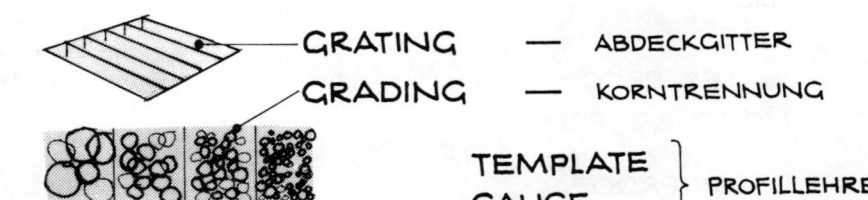

GRATING — ABDECKGITTER

GRADING — KORNTRENNUNG

TEMPLATE
GAUGE } PROFILLEHRE

CUBICLE   KABINE   E.G. TOILET CUBICLE   — WC-ZELLE
BOOTH   ABTEIL   E.G. TELLER'S BOOTH — KASSIERERKABINE
                  ZELLE   E.G. TELEPHONE BOX   — TELEFONZELLE

PAMPHLET
LEAFLET   } PROSPEKT
PROSPECTUS

FIXING INSTRUCTIONS — EINBAUANWEISUNG

BAUSTELLE   BUILDING SITE

BUILDING SITE
BAUSTELLE, BAUGELÄNDE

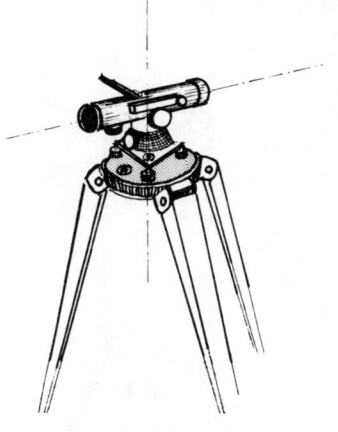

SURVEY OF SITE
ABMESSEN DER BAUSTELLE

LEVELLING OF SITE
NIVELLIEREN DER BAUSTELLE

SIGHT RAIL
VISIERGERÜST

BATTER BOARD
SCHNURGERÜST

BUILDER'S LEVEL
NIVELLIERGERÄT

THEODOLITE — THEODOLIT

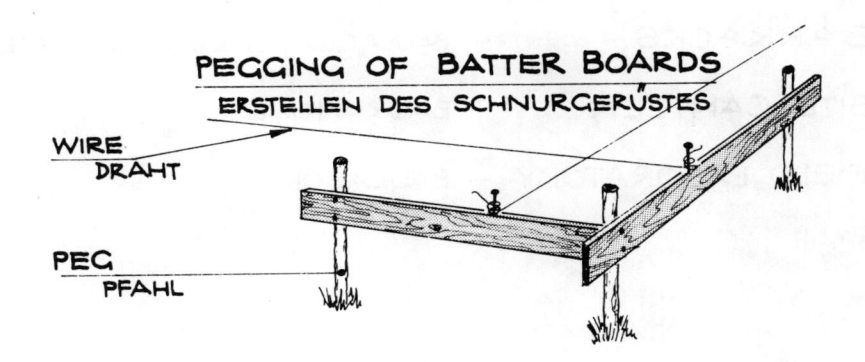

PEGGING OF BATTER BOARDS
ERSTELLEN DES SCHNURGERÜSTES

WIRE — DRAHT

PEG — PFAHL

BUILDING OPERATIONS
BUILDING WORKS   } BAUARBEITEN

SITE SUPERVISION   BAUSTELLENÜBERWACHUNG
                          BAULEITUNG

47

BAUSTELLE   BUILDING SITE

SITE INSTALLATIONS
BAUSTELLENEINRICHTUNGEN

SITE ROAD
BUILDER'S ROAD } BAUSTELLENSTRASSE

TEMPORARY WATER CONNECTION
PROVISORISCHER WASSERANSCHLUSS

STORAGE BIN, SILO
LAGERBEHÄLTER, SILO

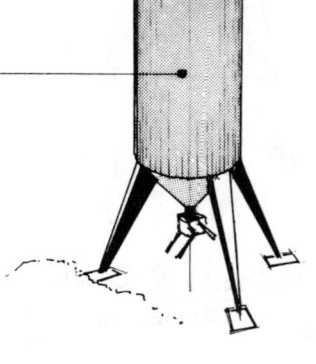

SITE OFFICE   — BAUBÜRO, BAUSTELLENBÜRO
BARRACKS   — BAULAGER, BAUUNTERKÜNFTE
SITE CANTEEN   — BAUKANTINE
FIELD LABORATORY   — BAULABOR

SITE CLEARANCE
CLEARING OF SITE } BAUSTELLENRÄUMUNG

BAUMASCHINEN

BUILDING MACHINERY

BUILDING MACHINERY — BAUMASCHINEN

Some Common Machines
Einige Gebräuchliche Maschinen

DROP HAMMER — Ramme, Fallbär

CRAWLER-LOADER CATERPILLAR — Raupe, Planierraupe

SHOVEL — Schaufel

EXCAVATOR — Bagger

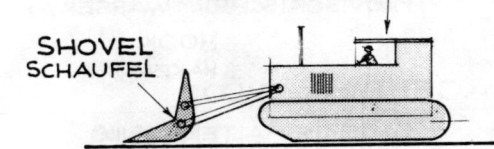

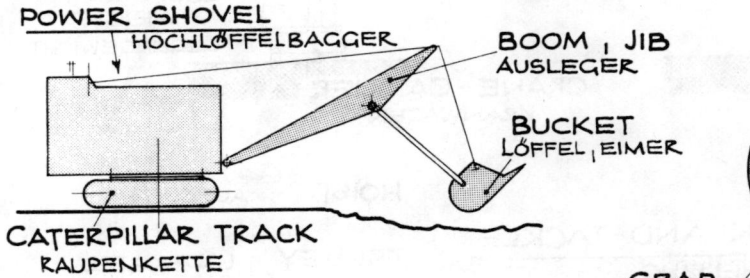

POWER SHOVEL — Hochlöffelbagger

BOOM, JIB — Ausleger

BUCKET — Löffel, Eimer

CATERPILLAR TRACK — Raupenkette

GRAB / GRABBING BUCKET — Greifer

BACKACTING SHOVEL, BACKHOE — Tieflöffelbagger

DREDGER — Schwimmbagger

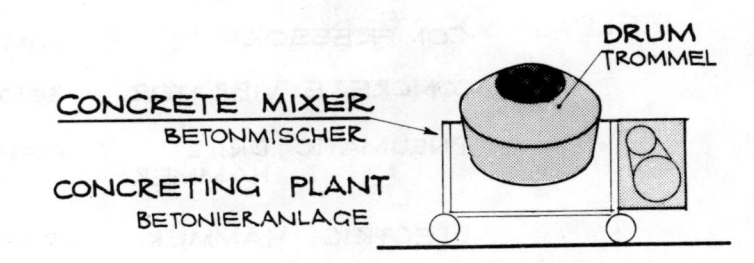

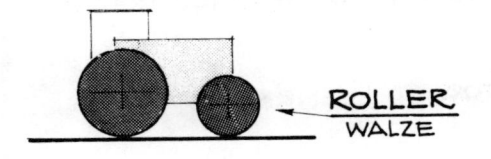

DRUM — Trommel

CONCRETE MIXER — Betonmischer

CONCRETING PLANT — Betonieranlage

ROLLER — Walze

PUMP — Pumpe

CENTRIFUGAL PUMP — Kreiselpumpe

GUNITING MACHINE — Torkretiermaschine

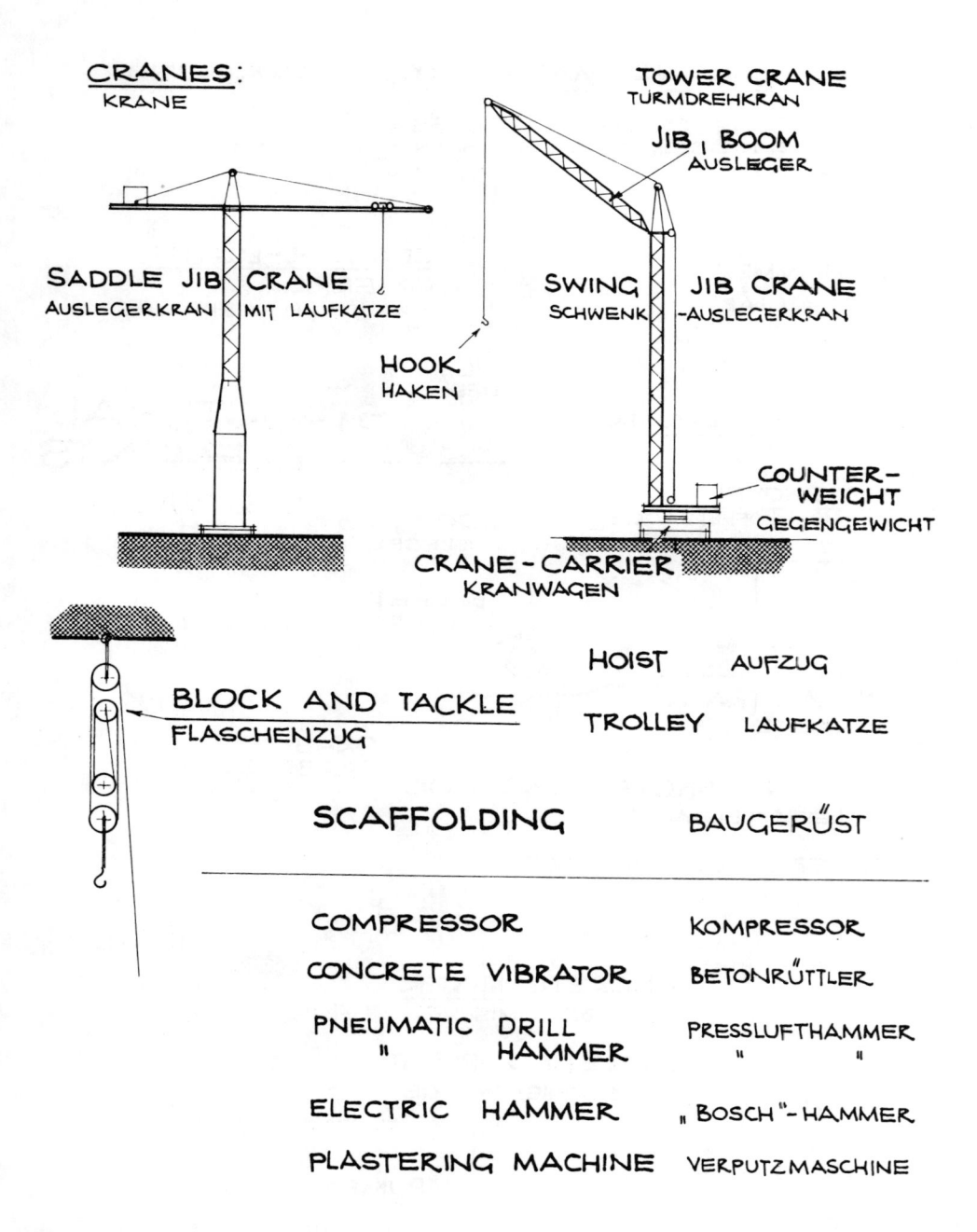

BAUMASCHINEN — BUILDING MACHINERY

CRANES:
KRANE

SADDLE JIB CRANE — AUSLEGERKRAN MIT LAUFKATZE

TOWER CRANE — TURMDREHKRAN

JIB, BOOM — AUSLEGER

SWING JIB CRANE — SCHWENK-AUSLEGERKRAN

HOOK — HAKEN

COUNTERWEIGHT — GEGENGEWICHT

CRANE-CARRIER — KRANWAGEN

BLOCK AND TACKLE — FLASCHENZUG

HOIST — AUFZUG

TROLLEY — LAUFKATZE

SCAFFOLDING — BAUGERÜST

COMPRESSOR	KOMPRESSOR
CONCRETE VIBRATOR	BETONRÜTTLER
PNEUMATIC DRILL	PRESSLUFTHAMMER
" HAMMER	" "
ELECTRIC HAMMER	„BOSCH"-HAMMER
PLASTERING MACHINE	VERPUTZMASCHINE

ENGINE	VERBRENNUNGSMOTOR
MOTOR	ELEKTROMOTOR

STRUCTURAL ELEMENTS
BAUTEILE

STRUCTURAL ELEMENTS
BAUTEILE, TRAGTEILE

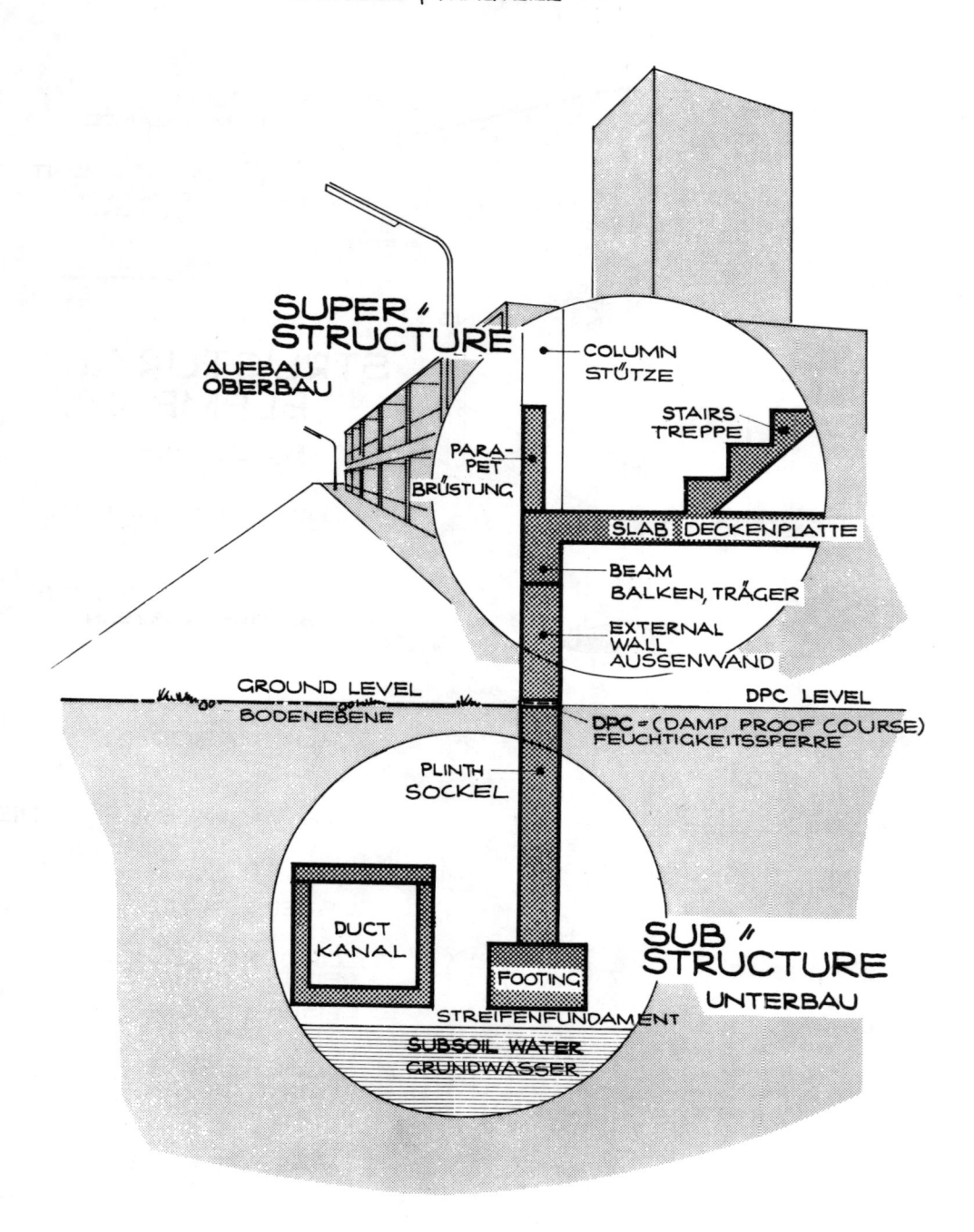

SUPER-STRUCTURE
AUFBAU, OBERBAU

- COLUMN / STÜTZE
- STAIRS / TREPPE
- PARAPET / BRÜSTUNG
- SLAB / DECKENPLATTE
- BEAM / BALKEN, TRÄGER
- EXTERNAL WALL / AUSSENWAND

GROUND LEVEL / BODENEBENE

DPC LEVEL
DPC = (DAMP PROOF COURSE) / FEUCHTIGKEITSSPERRE

- PLINTH / SOCKEL
- DUCT / KANAL
- FOOTING / STREIFENFUNDAMENT
- SUBSOIL WATER / GRUNDWASSER

SUB-STRUCTURE
UNTERBAU

UNTERBAU   **GROUND, TERRAIN**   SUBSTRUCTURE
Gelände, Grund, Boden

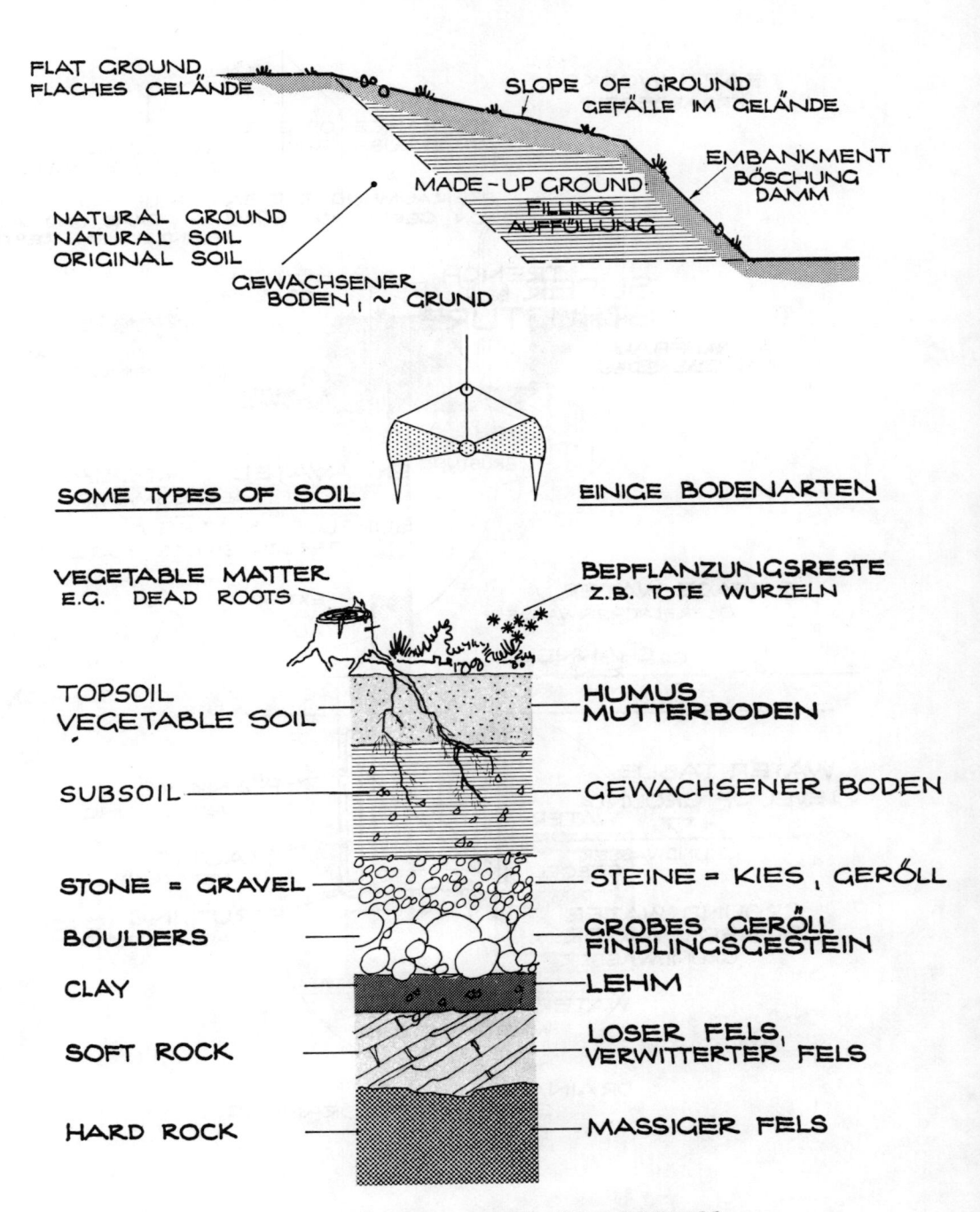

FLAT GROUND / FLACHES GELÄNDE	
SLOPE OF GROUND	GEFÄLLE IM GELÄNDE
EMBANKMENT	BÖSCHUNG / DAMM
MADE-UP GROUND / FILLING	AUFFÜLLUNG
NATURAL GROUND / NATURAL SOIL / ORIGINAL SOIL	GEWACHSENER BODEN, ~ GRUND

SOME TYPES OF SOIL   EINIGE BODENARTEN

VEGETABLE MATTER E.G. DEAD ROOTS	BEPFLANZUNGSRESTE Z.B. TOTE WURZELN
TOPSOIL / VEGETABLE SOIL	HUMUS / MUTTERBODEN
SUBSOIL	GEWACHSENER BODEN
STONE = GRAVEL	STEINE = KIES, GERÖLL
BOULDERS	GROBES GERÖLL / FINDLINGSGESTEIN
CLAY	LEHM
SOFT ROCK	LOSER FELS, VERWITTERTER FELS
HARD ROCK	MASSIGER FELS

E.G. COMMON GROUND INTERSPERSED WITH BOULDERS
Z.B. BINDIGER BODEN MIT FINDLINGEN DURCHSETZT

UNTERBAU ## EXCAVATION SUBSTRUCTURE
AUSHUB

EARTHWORKS
ERDARBEITEN

CLEARANCE OF SITE
OF SHRUBS, BUSHES, GRASS & ALL
VEGETABLE MATTER

FREIRÄUMUNG DER BAUSTELLE
VON GESTRÄUCH, BÜSCHEN, GRAS UND ALLEN
PFLANZLICHEN RESTEN

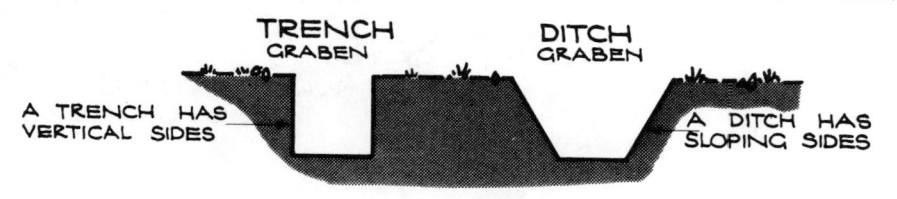

TRENCH **DITCH**
GRABEN   GRABEN

A TRENCH HAS   A DITCH HAS
VERTICAL SIDES   SLOPING SIDES

EXCAVATED MATERIAL
AUSHUBMATERIAL
SURPLUS MATERIAL
ÜBERSCHÜSSIGES MATERIAL

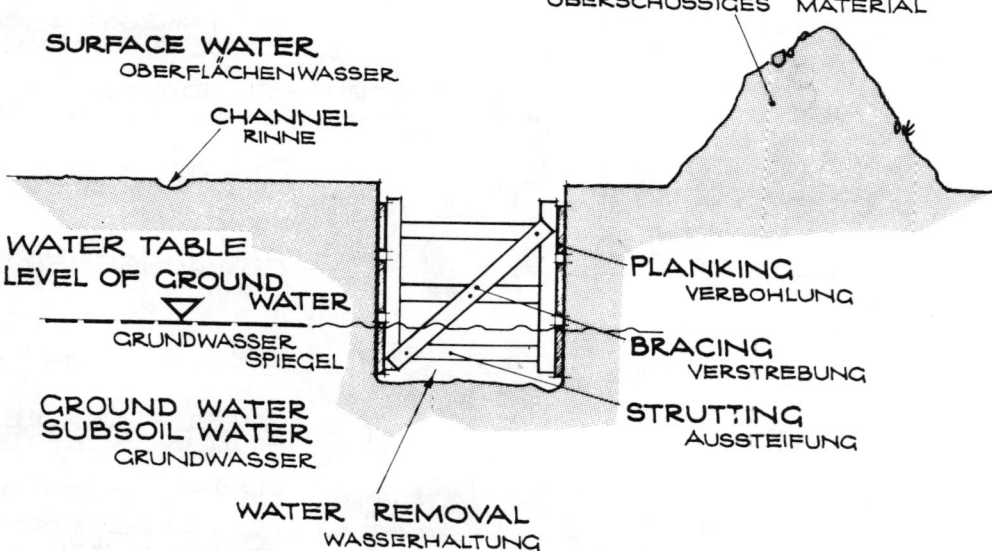

SURFACE WATER
OBERFLÄCHENWASSER

CHANNEL
RINNE

WATER TABLE
LEVEL OF GROUND WATER
GRUNDWASSER SPIEGEL

PLANKING
VERBOHLUNG

BRACING
VERSTREBUNG

GROUND WATER
SUBSOIL WATER
GRUNDWASSER

STRUTTING
AUSSTEIFUNG

WATER REMOVAL
WASSERHALTUNG

DRAINAGE
ENTWÄSSERUNG, DRÄNUNG

TO LEVEL   PLANIEREN

UNTERBAU **FOUNDATION** SUBSTRUCTURE
FUNDIERUNG, GRÜNDUNG
FUNDAMENT

FOOTING
  BANKETT, GRÜNDUNG

BASE
  EINZELFUNDAMENT
  (WÖRTLICH: BANKETT – JEDOCH UNGEBRÄUCHLICH)

WALL FOOTING
  MAUERFUNDAMENT

FOUNDATION SLAB
BASE SLAB
  FUNDAMENTPLATTE

MACHINE BASE
  MASCHINENFUNDAMENT

FLOOR SLAB
  BODENPLATTE

TIE BEAM
  ZERRBALKEN

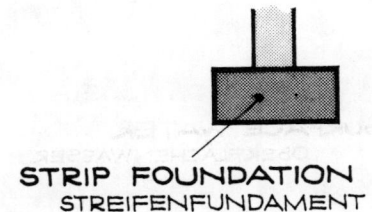

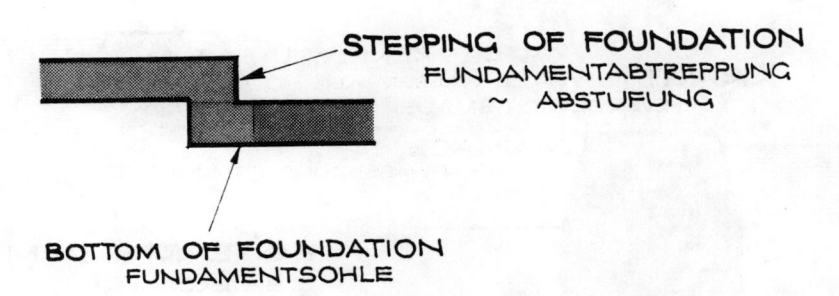

STRIP FOUNDATION
  STREIFENFUNDAMENT

STEPPING OF FOUNDATION
  FUNDAMENTABTREPPUNG
  ~ ABSTUFUNG

BOTTOM OF FOUNDATION
  FUNDAMENTSOHLE

UNTERBAU

PILING
PFAHLGRÜNDUNG

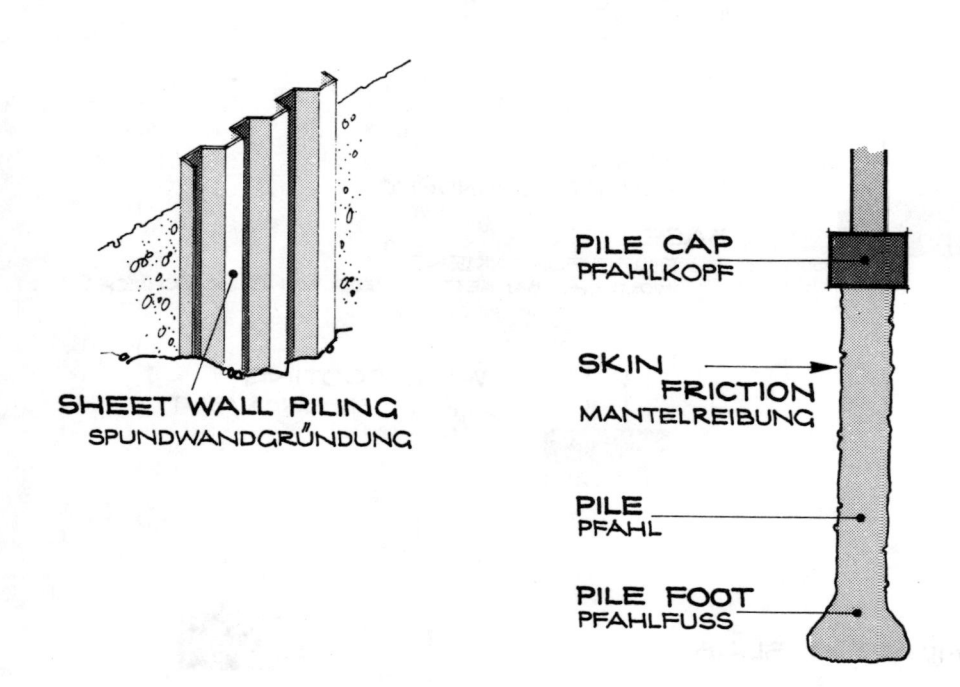

SUBSTRUCTURE

SHEET WALL PILING
SPUNDWANDGRÜNDUNG

PILE CAP — PFAHLKOPF

SKIN FRICTION — MANTELREIBUNG

PILE — PFAHL

PILE FOOT — PFAHLFUSS

RETAINING WALLS
STÜTZMAUERN

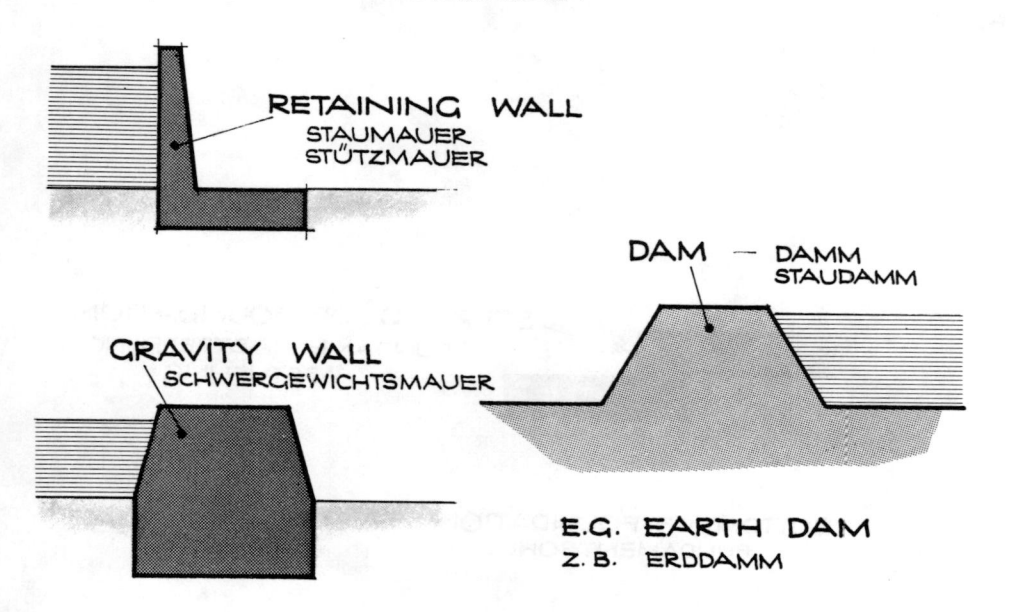

RETAINING WALL
STAUMAUER
STÜTZMAUER

GRAVITY WALL
SCHWERGEWICHTSMAUER

DAM — DAMM
STAUDAMM

E.G. EARTH DAM
Z. B. ERDDAMM

UNTERBAU **DUCTS** SUBSTRUCTURE
BODENKANÄLE

TUNNEL

NOTE: UNDERGROUND PARKING
TIEFGARAGE

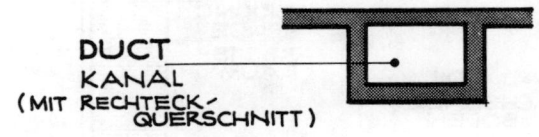

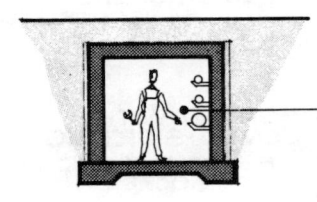

DUCT
KANAL
(MIT RECHTECK-
QUERSCHNITT)

ACCESSIBLE DUCT
BEGEHBARER LEITUNGSKANAL

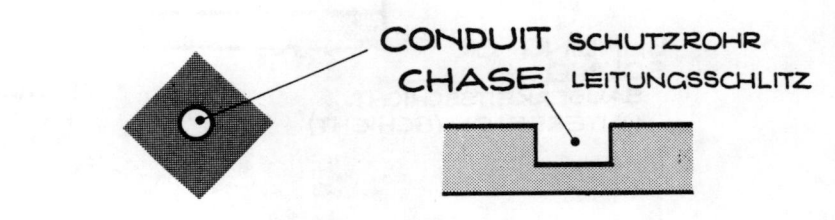

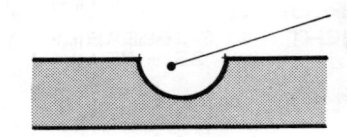

GRATING OR SLABS
ABDECKROST ODER - PLATTEN

OPEN FLOOR DUCT
OFFENER BODENKANAL

CONDUIT SCHUTZROHR
CHASE LEITUNGSSCHLITZ

HALFROUND FLOOR
CHANNEL
HALBKREISFÖRMIGE
BODENRINNE

GROOVE
RILLE, NUT

UNTERBAU   SUBSTRUCTURE

EXAMPLE:
SECTION THROUGH FOUNDATION
FUNDAMENTSCHNITT

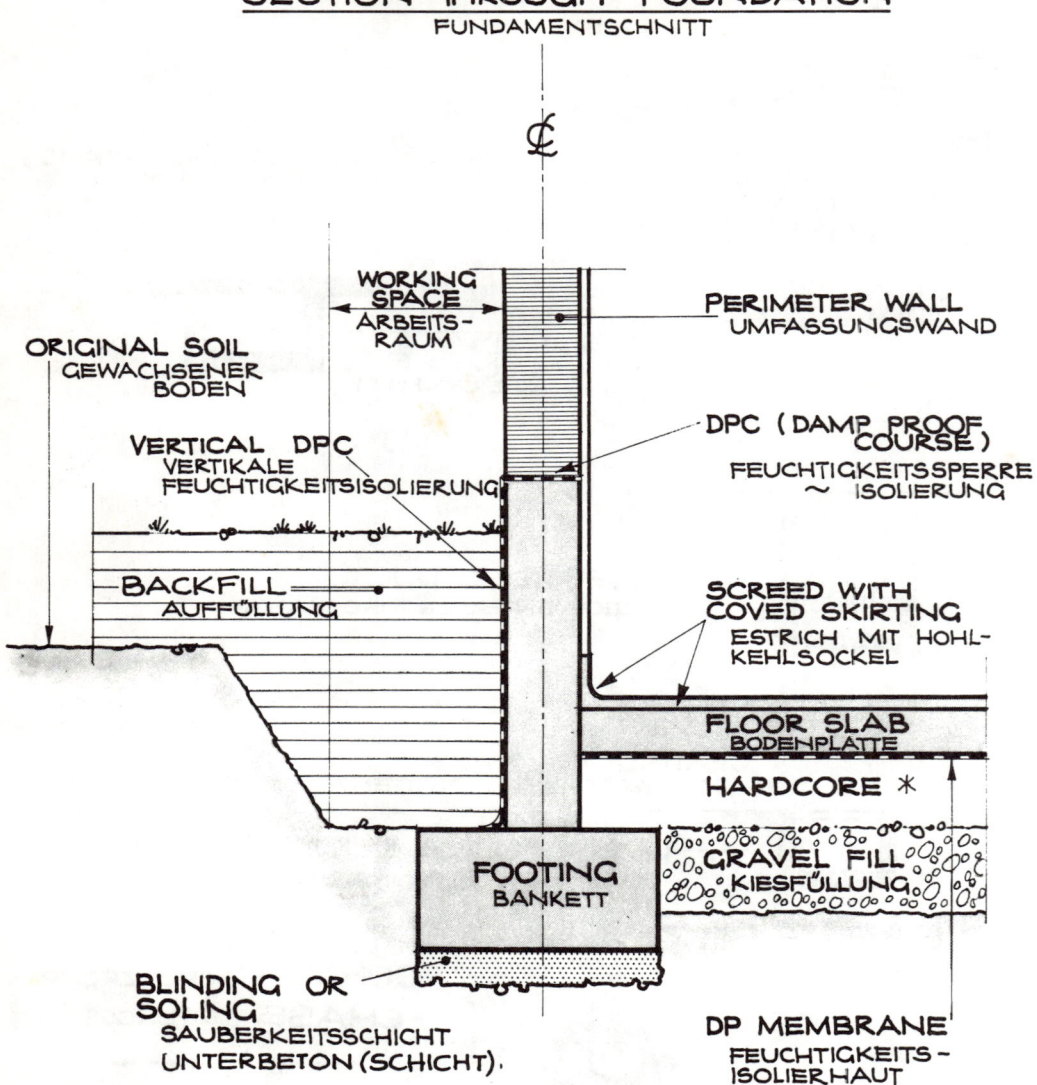

- **WORKING SPACE** — ARBEITSRAUM
- **PERIMETER WALL** — UMFASSUNGSWAND
- **ORIGINAL SOIL** — GEWACHSENER BODEN
- **DPC (DAMP PROOF COURSE)** — FEUCHTIGKEITSSPERRE ~ ISOLIERUNG
- **VERTICAL DPC** — VERTIKALE FEUCHTIGKEITSISOLIERUNG
- **BACKFILL** — AUFFÜLLUNG
- **SCREED WITH COVED SKIRTING** — ESTRICH MIT HOHLKEHLSOCKEL
- **FLOOR SLAB** — BODENPLATTE
- **HARDCORE** *
- **GRAVEL FILL** — KIESFÜLLUNG
- **FOOTING** — BANKETT
- **BLINDING OR SOLING** — SAUBERKEITSSCHICHT UNTERBETON (SCHICHT)
- **DP MEMBRANE** — FEUCHTIGKEITSISOLIERHAUT

* HARDCORE = CONSOLIDATED LAYER OF GRAVEL (1" TO 3")
VERDICHTETE KIESSCHICHT (IM SIEBBEREICH 25 – 75 MM)

BLINDINGS:

BLINDING OF 50 MM HARDCORE
SAUBERKEITSSCHICHT AUS 50 MM KIESPACKUNG

BRICK SOLING LAID ON EDGE
SAUBERKEITSSCHICHT AUS STEHENDER ZIEGELSCHICHT

UNTERBAU ## SOME NOTES TO EXCAVATIONS   SUBSTRUCTURE
EINIGE BEMERKUNGEN ZU AUSHUBARBEITEN

THE BOTTOM OF EXCAVATION SHALL
BE INSPECTED AND APPROVED
BY THE ENGINEER
BEFORE CONCRETE IS PLACED

> DIE AUSHUBSOHLE MUSS VOR DEM
> BETONIEREN VOM STATIKER
> INSPIZIERT UND ABGENOMMEN
> WERDEN

POCKETS OF SOFT GROUND SHALL
BE REMOVED AND THE RESULTING VOIDS
FILLED WITH APPROVED MATERIAL AND
CONSOLIDATED

> NESTER MIT SCHLECHTEM BAUGRUND
> SIND ZU ENTFERNEN, UND DIE SICH
> ERGEBENDEN HOHLRÄUME SIND
> MIT ABGENOMMENEM MATERIAL
> EINZUFÜLLEN UND ZU VERDICHTEN

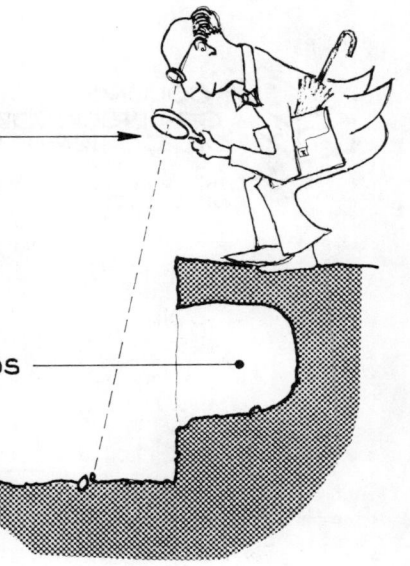

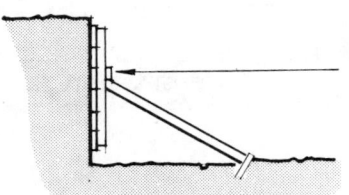

SIDES OF EXCAVATION SHALL BE
ADEQUATELY SUPPORTED BY PLANKING
AND STRUTTING TO PREVENT COLLAPSE

> DIE SEITEN DER AUSHUBGRÄBEN SIND
> DURCH ABBOLZUNG HINREICHEND VOR
> EINSTURZ ZU SICHERN

FILLING SHALL BE DEPOSITED IN LAYERS NOT EXCEEDING
200 MM THICK AND THOROUGHLY COMPACTED BY
APPROVED MEANS

> AUFFÜLLMATERIAL LAGENWEISE NICHT STÄRKER
> ALS 200 MM AUFTRAGEN UND NACH VORSCHRIFT
> VERDICHTEN

SURPLUS EXCAVATED MATERIAL SHALL BE DEPOSITED,
SPREAD AND LEVELLED ON SITE WHERE DIRECTED
OR REMOVED FROM THE SITE

> ÜBERSCHÜSSIGES AUSHUBMATERIAL AUFTRAGEN,
> VERTEILEN UND NACH ANGABE AUF DEM BAU-
> GELÄNDE PLANIEREN ODER ABFAHREN

OBERBAU ## COLUMNS SUPERSTRUCTURE
STÜTZEN

COLUMNS OF BRICKWORK OR CONCRETE
STÜTZEN AUS MAUERWERK ODER BETON

PILLAR
SÄULE

PIER
PFEILER

TAPER / ANLAUF

BUTTRESS
STREBEPFEILER

STANCHION
STAHLSTÜTZEN

OUTSIDE ⌀ / AUSSEN ⌀

TUBE
ROHRSTÜTZE
E.G. TUBULAR SCAFFOLDING — ROHR-GERÜST

MULLION
LEICHTE STAHLSTÜTZE
(BIS CA. 200/200 mm)

I-BEAM
SCHWERE STAHLSTÜTZE

COLUMN OF COMPOUND SECTIONS
ZUSAMMENGESETZTE PROFILSTÜTZE

COLUMNS OF TIMBER
HOLZSTÜTZEN

PROP
ABBOLZUNGSSTÜTZE

POST
PFOSTEN

POLE
MAST

OBERBAU   **WALLS**   SUPERSTRUCTURE
MAUERN, WÄNDE

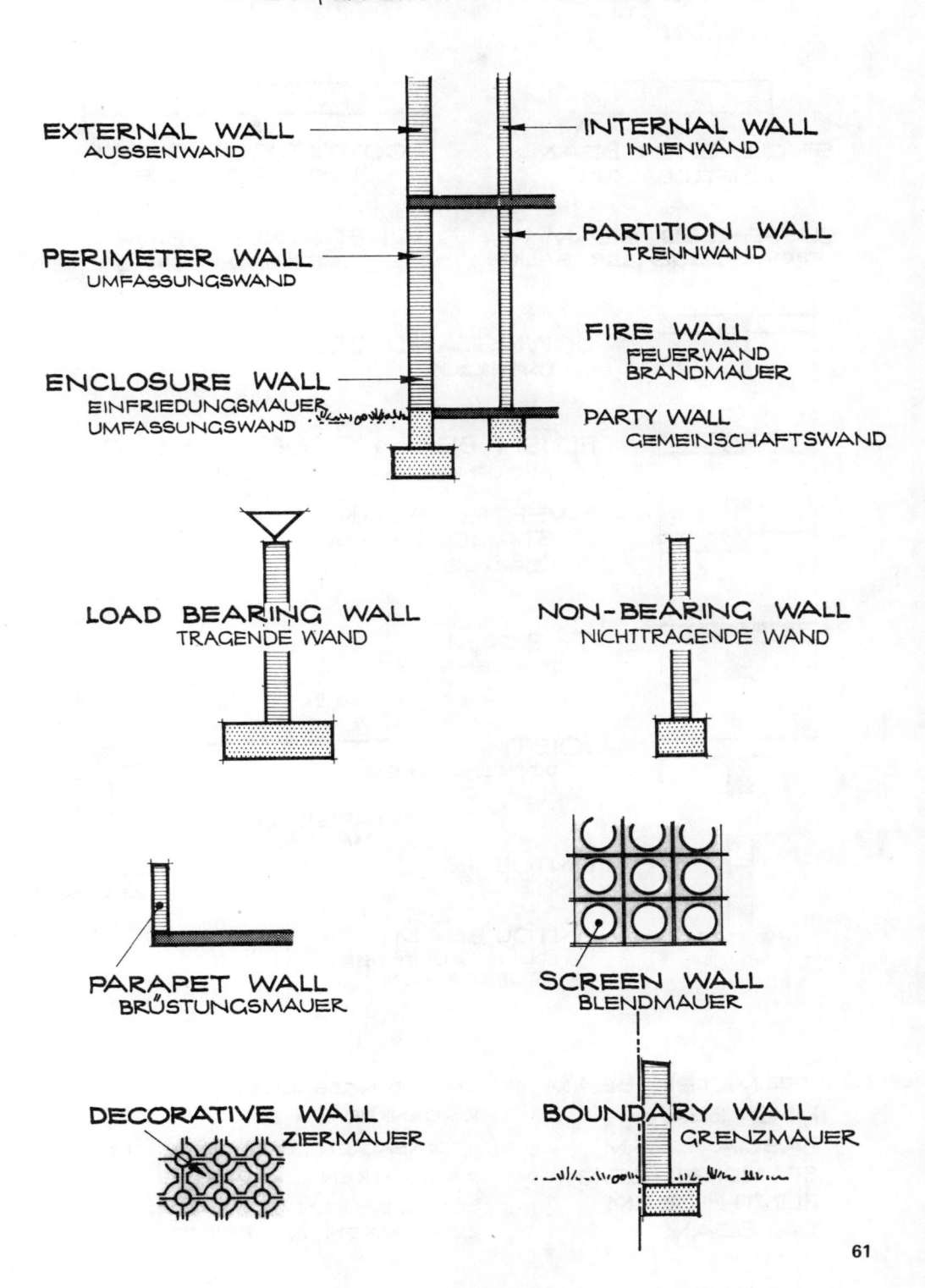

EXTERNAL WALL
AUSSENWAND

INTERNAL WALL
INNENWAND

PERIMETER WALL
UMFASSUNGSWAND

PARTITION WALL
TRENNWAND

FIRE WALL
FEUERWAND
BRANDMAUER

ENCLOSURE WALL
EINFRIEDUNGSMAUER
UMFASSUNGSWAND

PARTY WALL
GEMEINSCHAFTSWAND

LOAD BEARING WALL
TRAGENDE WAND

NON-BEARING WALL
NICHTTRAGENDE WAND

PARAPET WALL
BRÜSTUNGSMAUER

SCREEN WALL
BLENDMAUER

DECORATIVE WALL
ZIERMAUER

BOUNDARY WALL
GRENZMAUER

61

OBERBAU   **BEAMS**   SUPERSTRUCTURE
BALKEN, RIEGEL, UNTERZÜGE
TRÄGER

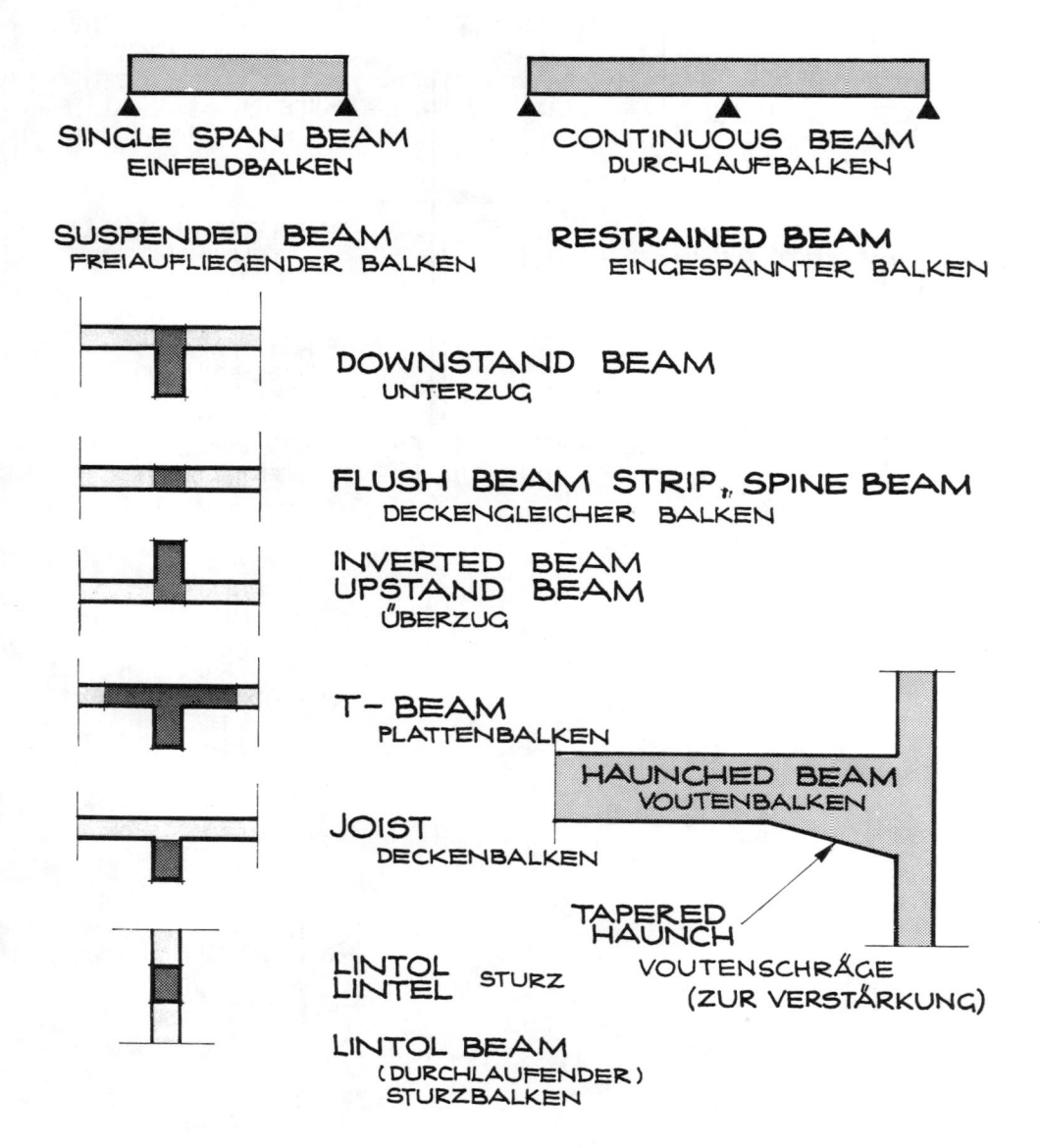

SINGLE SPAN BEAM
EINFELDBALKEN

CONTINUOUS BEAM
DURCHLAUFBALKEN

SUSPENDED BEAM
FREIAUFLIEGENDER BALKEN

RESTRAINED BEAM
EINGESPANNTER BALKEN

DOWNSTAND BEAM
UNTERZUG

FLUSH BEAM STRIP, SPINE BEAM
DECKENGLEICHER BALKEN

INVERTED BEAM
UPSTAND BEAM
ÜBERZUG

T-BEAM
PLATTENBALKEN

HAUNCHED BEAM
VOUTENBALKEN

JOIST
DECKENBALKEN

TAPERED HAUNCH
VOUTENSCHRÄGE
(ZUR VERSTÄRKUNG)

LINTOL
LINTEL   STURZ

LINTOL BEAM
(DURCHLAUFENDER)
STURZBALKEN

PERIMETER BEAM	UMFASSUNGSBALKEN
RING BEAM	RINGANKER
FASCIA BEAM	ATTIKABALKEN, FASSADENBALKEN
SPANDREL BEAM	RANDBALKEN
PLINTH BEAM	SOCKELBALKEN
TIE BEAM	ZERRBALKEN, QUERRIEGEL

OBERBAU   **SLABS**   SUPERSTRUCTURE
DECKEN, PLATTEN

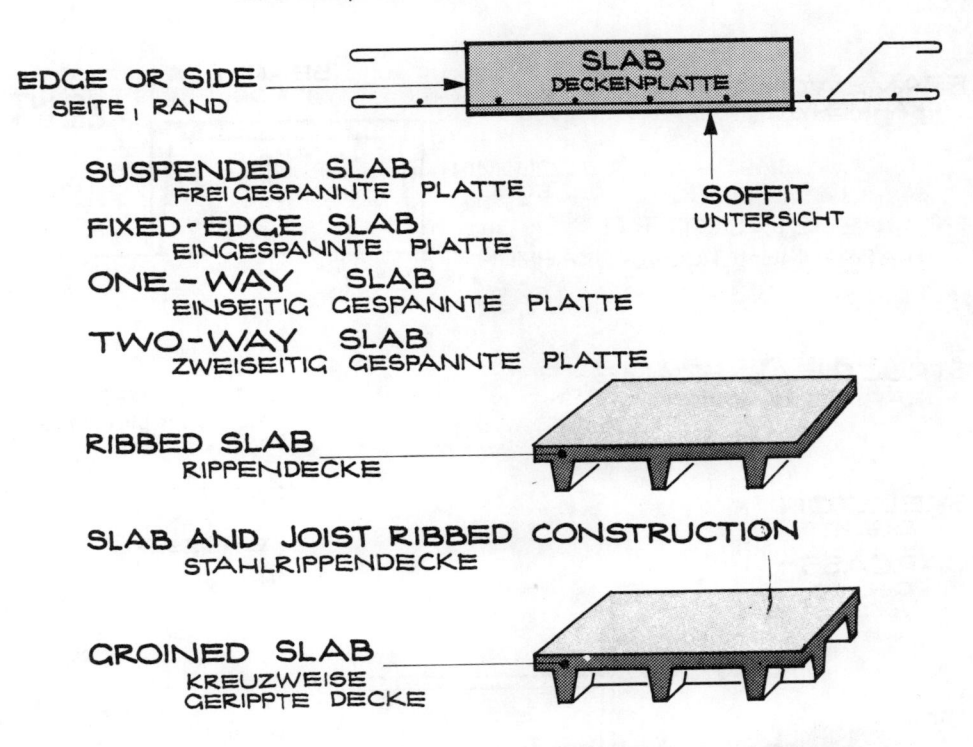

EDGE OR SIDE
SEITE, RAND

SLAB
DECKENPLATTE

SOFFIT
UNTERSICHT

SUSPENDED SLAB
FREIGESPANNTE PLATTE

FIXED-EDGE SLAB
EINGESPANNTE PLATTE

ONE-WAY SLAB
EINSEITIG GESPANNTE PLATTE

TWO-WAY SLAB
ZWEISEITIG GESPANNTE PLATTE

RIBBED SLAB
RIPPENDECKE

SLAB AND JOIST RIBBED CONSTRUCTION
STAHLRIPPENDECKE

GROINED SLAB
KREUZWEISE GERIPPTE DECKE

OPEN OR SMOOTH SOFFIT
OFFENE ODER GLATTE UNTERSICHT

MUSHROOM SLAB
PILZDECKE

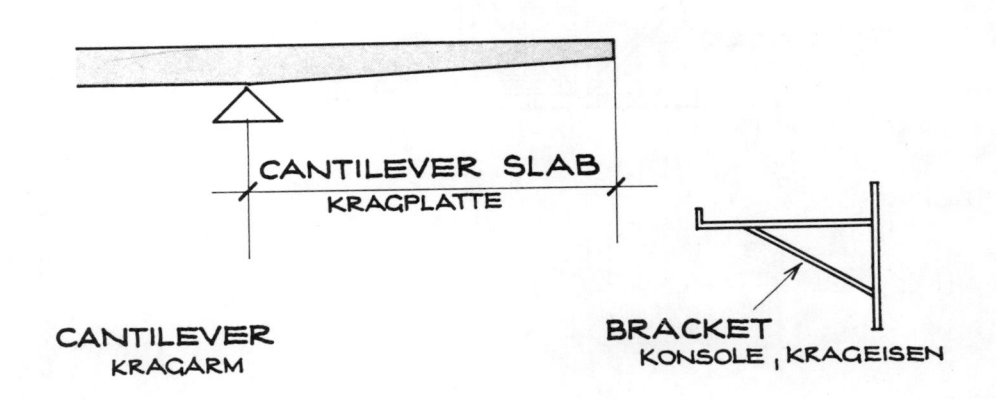

CANTILEVER SLAB
KRAGPLATTE

CANTILEVER
KRAGARM

BRACKET
KONSOLE, KRAGEISEN

OBERBAU   **FRAMES**   SUPERSTRUCTURE
RAHMEN

FRAMEWORK
RAHMENWERK

BEAM — RIEGEL
JOINT — GELENK
COLUMN — STIEL

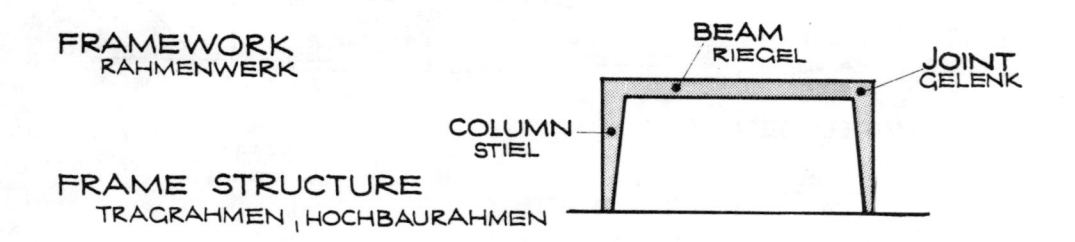

FRAME STRUCTURE
TRAGRAHMEN ǀ HOCHBAURAHMEN

STRUCTURAL FRAME
RAHMENTRAGWERK

SUSPENDED
FREI AUFLIEGEND

SKELETON
SKELETT

CARCASS
GERIPPE

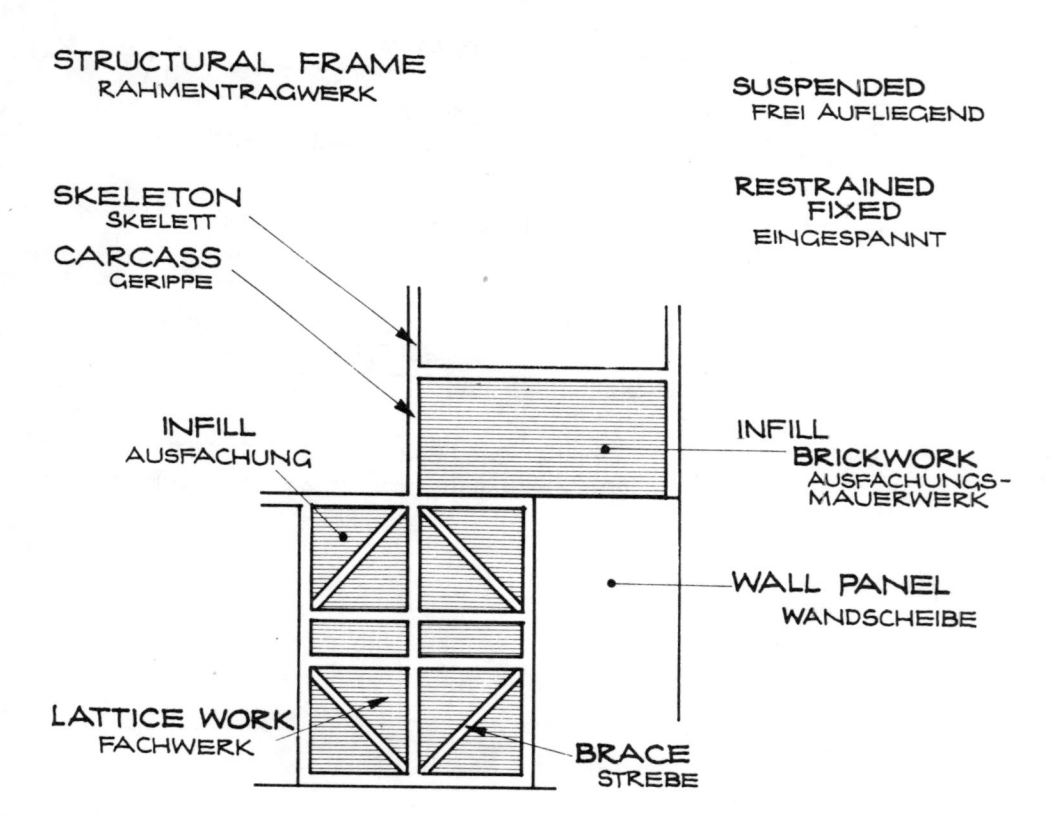

RESTRAINED FIXED
EINGESPANNT

INFILL
AUSFACHUNG

INFILL BRICKWORK
AUSFACHUNGS-MAUERWERK

WALL PANEL
WANDSCHEIBE

LATTICE WORK
FACHWERK

BRACE
STREBE

OBERBAU ## ROOF TYPES SUPERSTRUCTURE
DACHARTEN

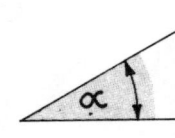

ROOF PITCH
DACHNEIGUNG
DACHGEFÄLLE

DOUBLE PITCH ROOF
GABLE ROOF
SATTELDACH

MONO PITCH ROOF
PULTDACH

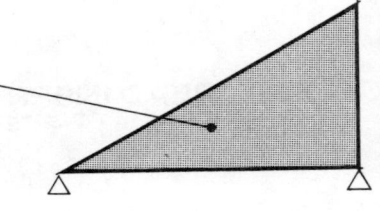

FLAT ROOF
FLACHDACH

CANOPY
(AUSKRAGENDES)
SCHUTZDACH

SAWTOOTH ROOF
SHEDDACH

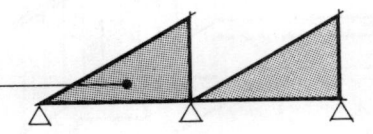

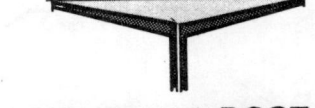

DOME
CUPOLA
KUPPEL

MUSHROOM ROOF
PILZDACH

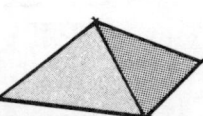

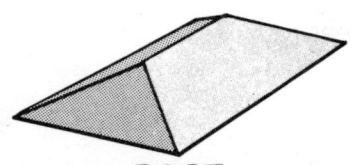

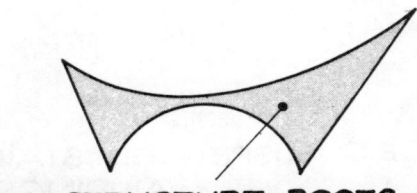

TENT ROOF
ZELTDACH

HIP ROOF
WALMDACH

SHELL STRUCTURE ROOFS
SCHALENDACHKONSTRUKTION

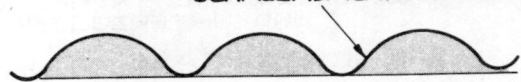

OBERBAU ## STAIRS AND STEPS SUPERSTRUCTURE
TREPPENSTUFEN UND EINZELSTUFEN

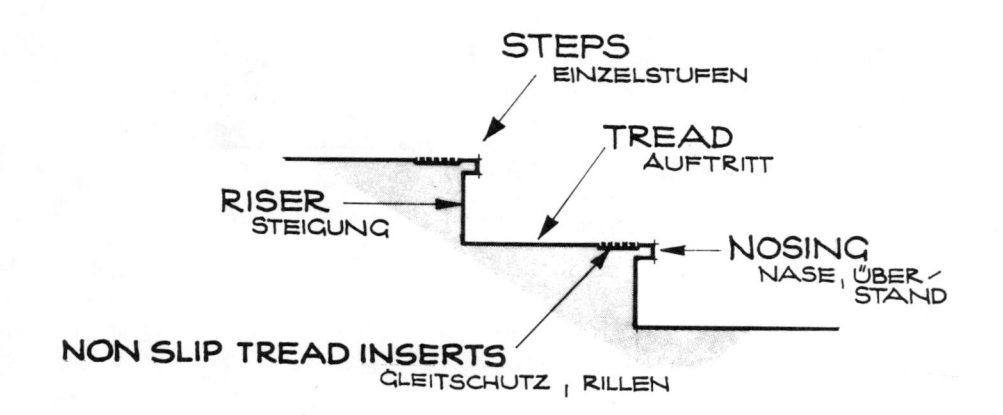

STEPS — EINZELSTUFEN
TREAD — AUFTRITT
RISER — STEIGUNG
NOSING — NASE, ÜBERSTAND
NON SLIP TREAD INSERTS — GLEITSCHUTZ, RILLEN

STAIRCASE
TREPPE, TREPPENHAUS

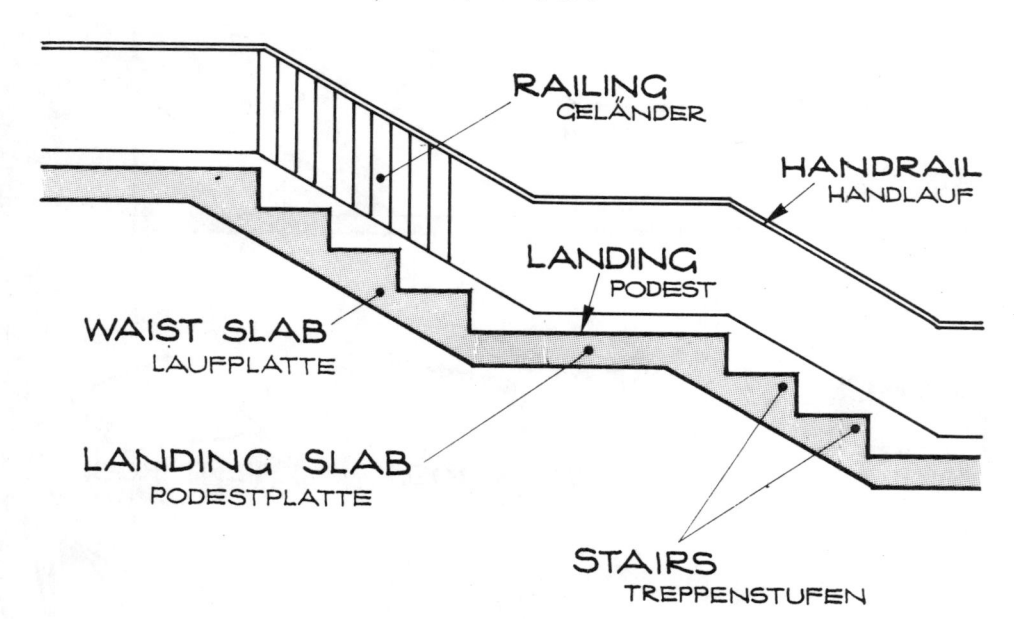

RAILING — GELÄNDER
HANDRAIL — HANDLAUF
LANDING — PODEST
WAIST SLAB — LAUFPLATTE
LANDING SLAB — PODESTPLATTE
STAIRS — TREPPENSTUFEN

E.G. SUSPENDED STAIRCASE WITH LANDINGS AND BEAMS
FREIAUFLIEGENDE TREPPE MIT PODESTEN UND UNTERZÜGEN

SPIRAL STAIRCASE   WENDELTREPPE

OBERBAU

ROOF MEMBERS
TEILE EINES DACHES

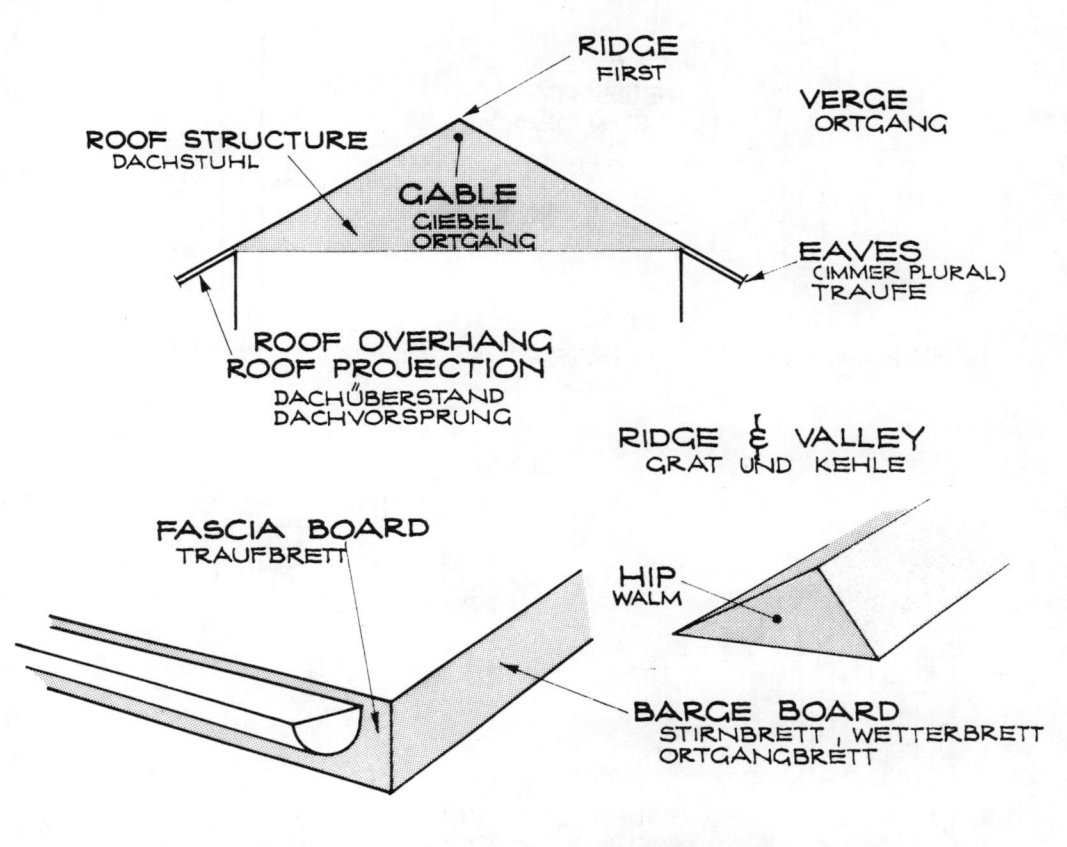

SUPERSTRUCTURE

RIDGE
FIRST

ROOF STRUCTURE
DACHSTUHL

VERGE
ORTGANG

GABLE
GIEBEL
ORTGANG

EAVES
(IMMER PLURAL)
TRAUFE

ROOF OVERHANG
ROOF PROJECTION
DACHÜBERSTAND
DACHVORSPRUNG

RIDGE & VALLEY
GRAT UND KEHLE

FASCIA BOARD
TRAUFBRETT

HIP
WALM

BARGE BOARD
STIRNBRETT, WETTERBRETT
ORTGANGBRETT

FASCIA = ATTIKA (UMLAUFENDER DACHRAND
BEIM FLACHDACH)

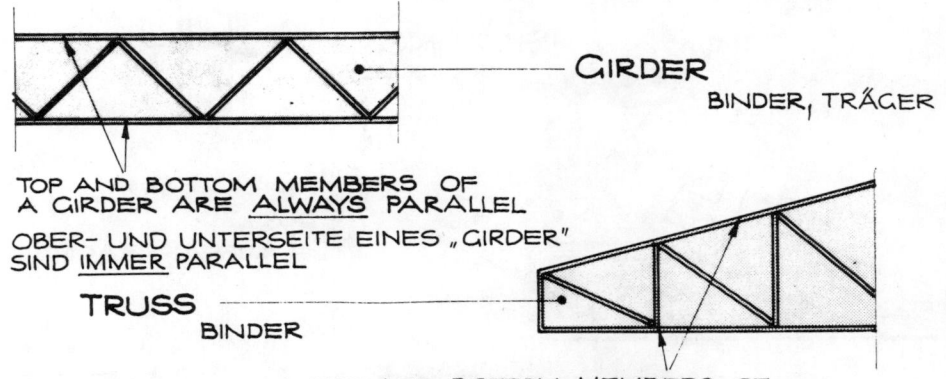

GIRDER
BINDER, TRÄGER

TOP AND BOTTOM MEMBERS OF
A GIRDER ARE ALWAYS PARALLEL

OBER- UND UNTERSEITE EINES „GIRDER"
SIND IMMER PARALLEL

TRUSS
BINDER

TOP AND BOTTOM MEMBERS OF
A TRUSS ARE NOT PARALLEL

OBER- UND UNTERSEITE EINES „TRUSS"
SIND NIE PARALLEL

OBERBAU · SUPERSTRUCTURE

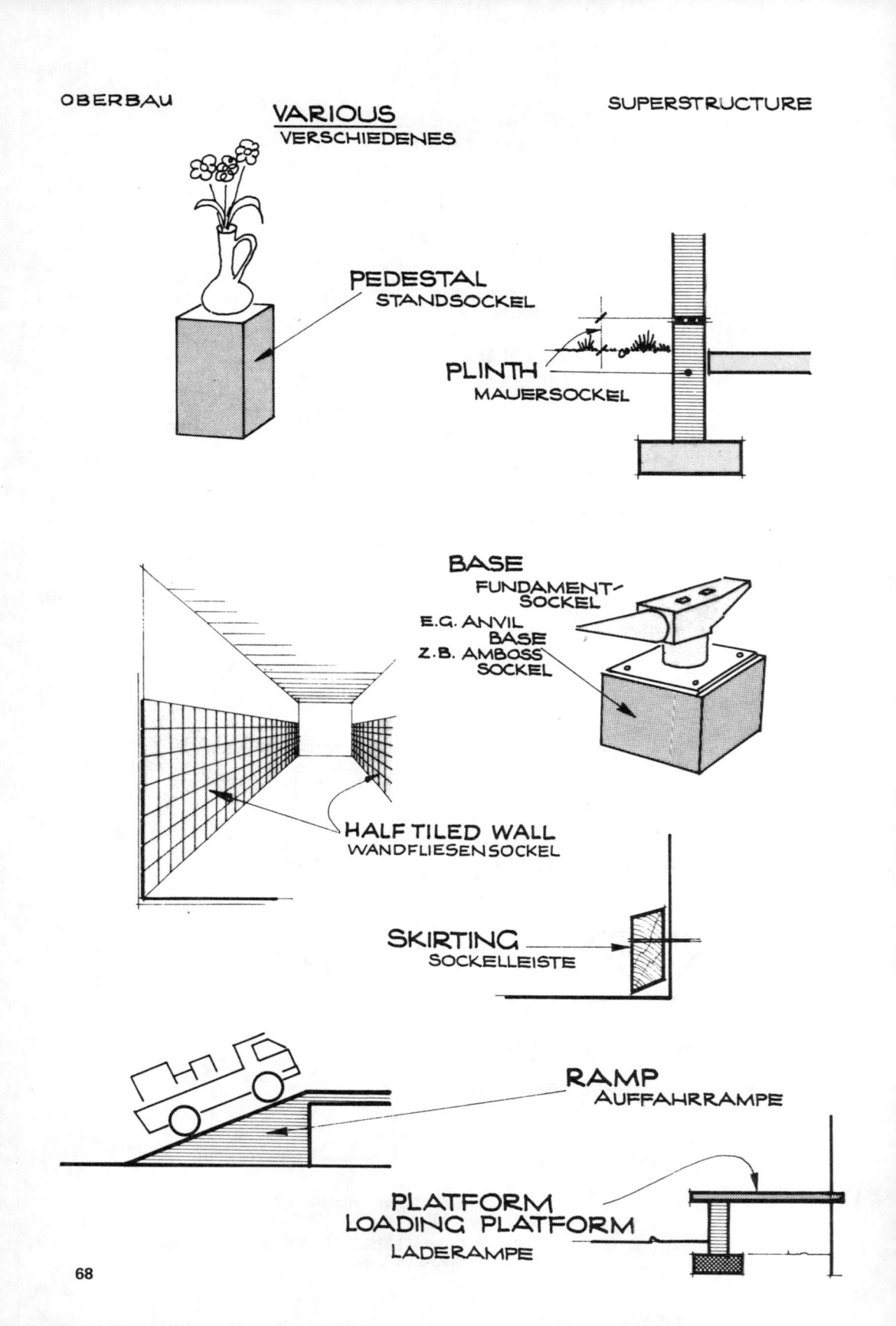

VARIOUS
VERSCHIEDENES

PEDESTAL
STANDSOCKEL

PLINTH
MAUERSOCKEL

BASE
FUNDAMENTSOCKEL
E.G. ANVIL BASE
Z.B. AMBOSS SOCKEL

HALF TILED WALL
WANDFLIESENSOCKEL

SKIRTING
SOCKELLEISTE

RAMP
AUFFAHRRAMPE

PLATFORM
LOADING PLATFORM
LADERAMPE

CONCRETE WORK
BETONARBEITEN

CONCRETE WORK
BETONARBEITEN

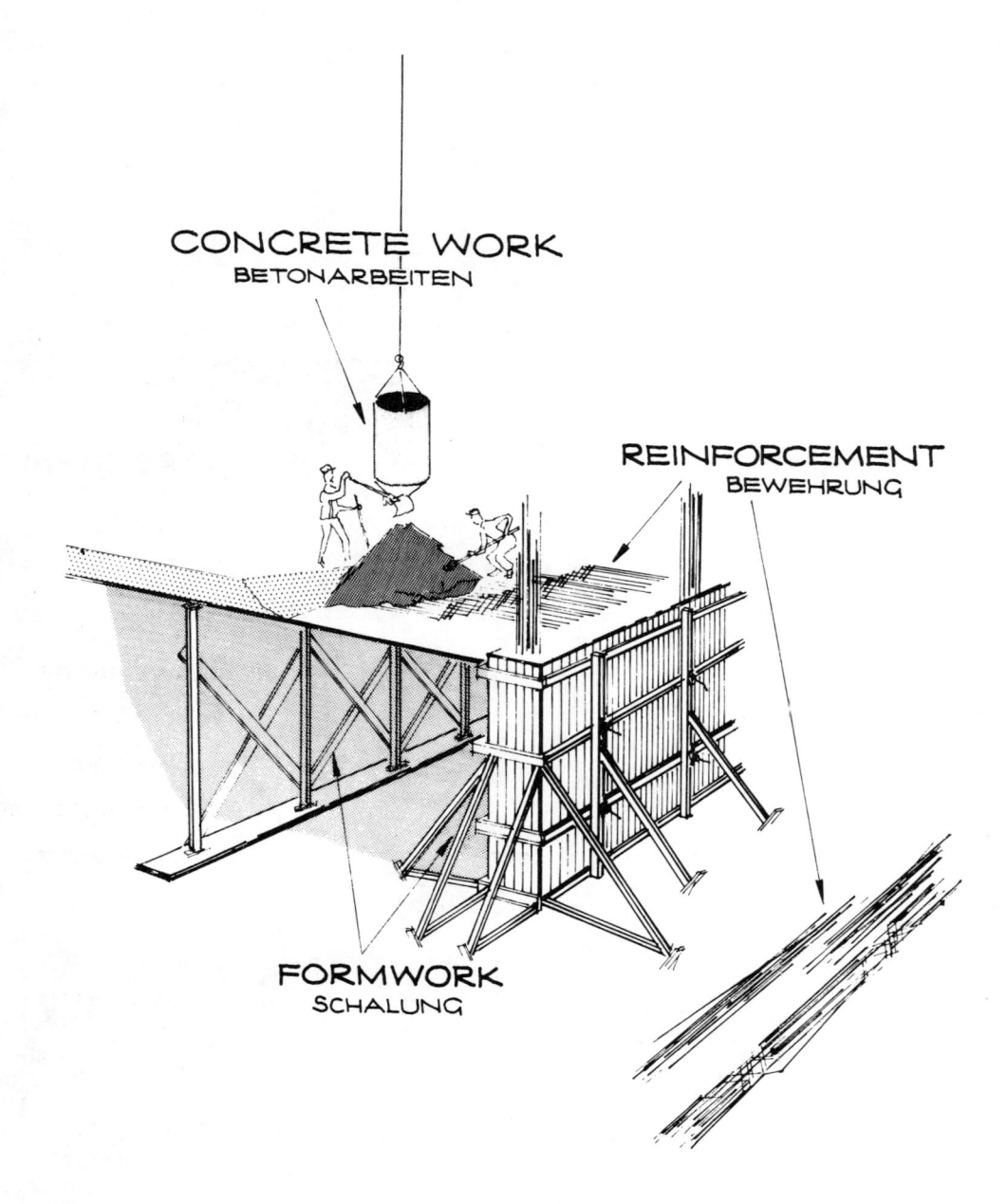

REINFORCEMENT
BEWEHRUNG

FORMWORK
SCHALUNG

70

BETONARBEITEN — **TYPES OF CONCRETE** — CONCRETE WORK
BETONARTEN

DIFFERENTIATION BY <u>MANUFACTURE</u>:   UNTERSCHEIDUNG IN DER <u>HERSTELLUNG</u>:

CAST-IN-SITU CONCRETE / IN-SITU CONCRETE (IN-SITU = IN SITUATION)	ORTBETON
READY-MIXED CONCRETE	TRANSPORTBETON
PRECAST CONCRETE	FERTIGBETON
PRESTRESSED CONCRETE	SPANNBETON
EXPOSED CONCRETE	SICHTBETON

BY <u>COMPONENTS USED</u>:   NACH BESTANDTEILEN:

LEAN CONCRETE	UNBEWEHRTER BETON / MAGERBETON
REINFORCED CONCRETE	STAHLBETON
SLAG CONCRETE	SCHLACKENBETON
SHINGLE CONCRETE	EINKORNBETON

BY <u>FUNCTION</u>:   NACH DER AUFGABE:

MASS CONCRETE	MASSENBETON, UNBEWEHRTER BETON
BREEZE CONCRETE	GEFÄLLEBETON
FOUNDATION CONCRETE	FUNDAMENTBETON

71

BETONARBEITEN COMPONENTS CONCRETE WORK
BESTANDTEILE

FINE AGGREGATE — FEINE ZUSCHLAGSTOFFE
COARSE AGGREGATE — GROBE ZUSCHLAGSTOFFE
CEMENT — ZEMENT
WATER — WASSER

TYPES OF CEMENT

NATURAL CEMENT — NATUR-ZEMENT
STANDARD SPECIFICATION CEMENT
SPECIFICATION CEMENT — NORMEN-ZEMENT

SLAG CEMENT
SLAGMENT — SCHLACKEN-ZEMENT

PORTLAND CEMENT — PORTLANDZEMENT

ORDINARY PORTLAND CEMENT
  GEWÖHNLICHER PORTLANDZEMENT

HIGH EARLY STRENGTH PORTLAND CEMENT
  HOCHWERTIGER PORTLANDZEMENT

RAPID HARDENING PORTLAND CEMENT
  SCHNELLERHÄRTENDER PORTLANDZEMENT

IRON PORTLAND CEMENT
  EISENPORTLANDZEMENT

BLAST FURNACE PORTLAND CEMENT
  HOCHOFEN PORTLANDZEMENT

NOTES: CEMENT SHALL BE PORTLAND CEMENT NORMAL SETTING TO STANDARD SPECIFICATION
  ZEMENT SOLLTE (IMMER) NORMAL ERSTARRENDER ODER GENORMTER PORTLANDZEMENT SEIN

STORAGE OF CEMENT IN A WEATHER-PROOF BUILDING ON A RAISED IMPERVIOUS FLOOR
  ZEMENTLAGERUNG IN EINEM WETTERGESCHÜTZEN GEBÄUDE IN EINEM NICHT ZUGÄNGLICHEN OBEREN STOCKWERK

BETONARBEITEN — **AGGREGATE** — CONCRETE WORK
ZUSCHLAGSTOFF

GRADING   KORNTRENNUNG
KORNGRÖSSE

FINE AGGREGATE (SAND)

WASHED NATURAL RIVER SAND
GEWASCHENER FLUSS-SAND

SAND DERIVED FROM CRUSHING GRAVEL
SAND AUS GEBROCHENEM KIES

PIT SAND   GRUBENSAND

NOTES: SAND SHALL BE EVENLY GRADED FROM FINE TO COARSE PARTICLES TO MAXIMUM 3/16" (5 mm) OF WHICH

BEACHTE: SAND SOLL GLEICHMÄSSIG ABGESTUFT SEIN VOM KLEINSTKORN ZUM GRÖSSTKORN VON MAXIMAL 5MM, DAVON SOLLTEN

10 – 30% SHOULD PASS THROUGH A SIEVE OF 52 MESHES PER SQUARE INCH

10 – 30% DURCH EIN SIEB MIT 52 MASCHEN PRO INCH^2 GEHEN

AND NOT OVER 10% SHOULD PASS THROUGH A SIEVE OF 100 MESHES PER S.I.

UND NICHT MEHR ALS 10% SOLLTEN DURCH EIN SIEB MIT 100 MASCHEN PRO INCH^2 GEHEN

SAND SHALL BE GRITTY, HARD PARTICLES FREE FROM DUST, CLAY, ANIMAL, VEGETABLE OR OTHER ORGANIC MATTER

SAND SOLL AUS RAUHEN, FESTEN KÖRNERN SEIN, FREI VON STAUB, TON, TIERISCHEN, PFLANZLISCHEN ODER ANDEREN ORGANISCHEN STOFFEN

COARSE AGGREGATE (STONE)

STONE	GESTEIN
GRAVEL	NATÜRLICHER KIES
RUBBLE	ROLLKIES
CHIPS	GEBROCHENER KIES
SHINGLES	SANDLOSER KIES
PEBBLES	KIESELSTEINE

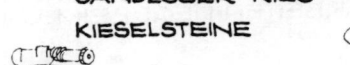

NOTES: COARSE AGGREGATE SHALL BE HARD, SOUND, DURABLE AND CLEAN STONE IN GRADUATIONS FROM 3/16" TO 2" Ø

GROBE ZUSCHLAGSTOFFE SOLLEN FESTE, EINWANDFREIE, DAUERHAFTE UND SAUBERE STEINE SEIN IN ABSTUFUNGEN VON 3/16" BIS 2" Ø

ALL STONE SHALL BE FREE FROM SAND, DUST, SALT, LIME, BITUMEN, CLAY OR OTHER DELETERIOUS MATTER

ALLE STEINE SOLLEN FREI VON SAND, STAUB, SALZ, KALK, BITUMEN, TON ODER ANDEREN SCHÄDLICHEN STOFFEN SEIN

BETONARBEITEN
MIXING OF CONCRETE
BETONMISCHEN
CONCRETE WORK

BATCHING BESTIMMEN DES MISCHUNGS-
VERHÄLTNISSES
   VOLUME BATCHING ~ NACH VOLUMENANTEILEN
   WEIGHT BATCHING ~ NACH GEWICHTANTEILEN

CONSISTENCY KONSISTENZ
WATER - CEMENT RATIO WASSER - ZEMENTFAKTOR

CONCRETE STRENGTH AFTER 28 DAYS
   BETONSPANNUNG NACH 28 TAGEN

TEST CUBE PROBEWÜRFEL
SLUMP TEST SCHWIND-
VERSUCH

VOLUME LOSS DUE
TO MIXING ± 42 %
   RAUMVERLUST DURCH
   MISCHEN ≃ 42 %

WATER TIGHT CONCRETE BY ADDITIVE
   WASSERDICHTEN BETON DURCH ZUSATZMITTEL

NOTE: THE FINAL RESPONSIBILITY AS TO STRENGTH,
DURABILITY, DENSITY, IMPERMEABILITY AND STABILITY
OF ALL CONCRETE WORK WILL BE THE CONTRACTOR'S

   DIE LETZTE VERANTWORTUNG IN BEZUG AUF DRUCKFESTIGKEIT
   LEBENSDAUER, DICHTE, DICHTIGKEIT UND HALTBARKEIT ALLER
   BETONARBEITEN LIEGT BEIM UNTERNEHMER

COMPACTION
VERDICHTUNG

COMPACTION BY MECHANICAL VIBRATORS
   VERDICHTUNG MIT MECHANISCHEN RÜTTLERN

HAND COMPACTION
   HANDVERDICHTUNG,
   STAMPFEN

BETONARBEITEN — CONCRETE WORK

TYPES OF MECHANICAL VIBRATORS
ARTEN MECHANISCHER RÜTTLER

SURFACE VIBRATOR
OBERFLÄCHENRÜTTLER

VIBRATOR BEAM
RÜTTELBOHLE

INTERNAL VIBRATOR
INNENRÜTTLER

E.G. VIBRATOR CYLINDER WITH FLEXIBLE SHAFT
z.B. FLASCHENRÜTTLER MIT BIEGSAMER WELLE

FOOT COMPACTED
FUSSVERDICHTET

NOTE: ALL CONCRETE IS TO BE THOROUGHLY COMPACTED BY INTERNAL VIBRATORS TO ENSURE SOLIDITY

BEACHTE: JEDER BETON IST GRÜNDLICH MIT INNENRÜTTLERN ZU VERDICHTEN, UM FESTIGKEIT ZU ERREICHEN

CURING (MATURING)
ABBINDEBEHANDLUNG

MATURING HEAT — ABBINDEWÄRME

NOTE: ALL NEWLY PLACED CONCRETE IS TO BE KEPT DAMP FOR NOT LESS THAN 10 DAYS BY MEANS OF WET SACKS OR SPRINKLERS

BEACHTE: FRISCH AUFGEBRACHTER BETON IST MINDESTENS 10 TAGE DURCH NASSE PLANEN ODER BESPRÜHEN FEUCHT ZU HALTEN

MAKING GOOD (PATCHING)
AUSBESSERUNGSARBEITEN

NOTE: REMOVE POCKETS OF LOOSE GRAVEL AND FILL RESULTING VOIDS AND OTHER HOLES WITH APPROVED CEMENT MORTAR 1:3 FOR MASS CONCRETE AND CEMENT MORTAR 1:2 FOR FINE AND REINFORCED CONCRETE

BEACHTE: KIESNESTER ENTFERNEN UND VERBLIEBENE HOHLRÄUME UND ANDERE ÖFFNUNGEN AUSFÜLLEN MIT BEWÄHRTEM ZEMENTMÖRTEL 1:3 FÜR MASSENBETON UND ZEMENTMÖRTEL 1:2 FÜR FEINEN UND BEWEHRTEN BETON

REINFORCEMENT
BEWEHRUNG

R.C. = REINFORCED CONCRETE = STAHLBETON
    E.G. R.C. SLAB, R.C. DESIGN, R.C. STRUCTURE

REINFORCEMENT BAR
REINFORCEMENT ROD   } BEWEHRUNGSSTAHL

REINFORCEMENT CAGE   BEWEHRUNGSKORB

MESH REINFORCEMENT   NETZBEWEHRUNG
STEEL FABRIC   BAUSTAHLGEWEBE (BStG)

TENSILE REINFORCEMENT   ZUGBEWEHRUNG
COMPRESSIVE REINFORCEMENT   DRUCKBEWEHRUNG
SHEAR REINFORCEMENT   SCHUBBEWEHRUNG

CONCRETE COVER FOR REINFORCEMENT
BETONÜBERDECKUNG DER BEWEHRUNG

CLEAR COVER IS TO BE PROVIDED 25 mm EVERYWHERE

ERFORDERLICHE ÜBERDECKUNG 25 mm ÜBERALL

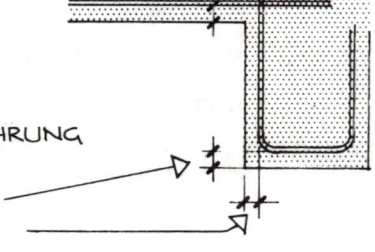

NOTE: IN NO CASE SHALL THE CONCRETE COVER BE LESS THAN THE DIAMETER OF THE ROD TO BE COVERED

BEACHTE: IN KEINEM FALL SOLL DIE BETONÜBERDECKUNG WENIGER ALS DER QUERSCHNITT DES ZU BEDECKENDEN STAHLS BETRAGEN

TYPES OF REINFORCEMENT BARS
ARTEN VON STAHLEINLAGEN

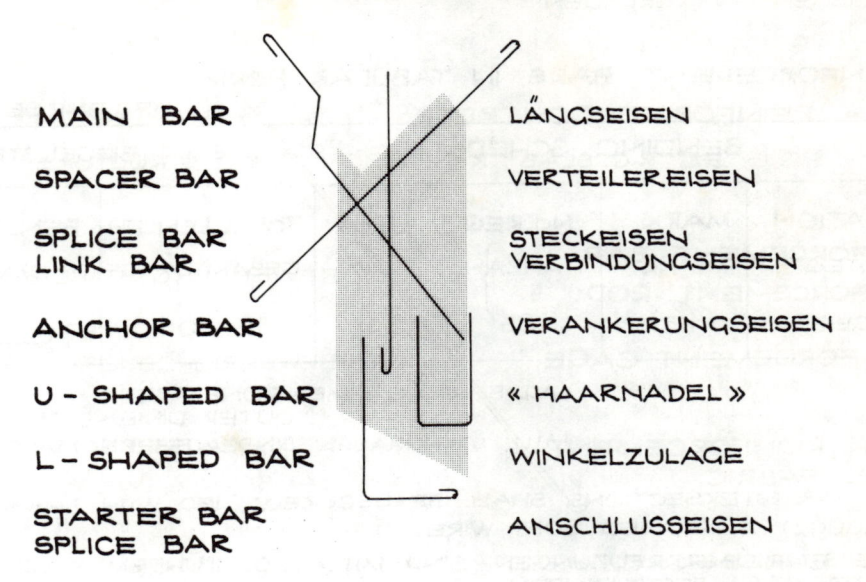

REINFORCEMENT
BEWEHRUNG

MAIN BAR	LÄNGSEISEN
SPACER BAR	VERTEILEREISEN
SPLICE BAR	STECKEISEN
LINK BAR	VERBINDUNGSEISEN
ANCHOR BAR	VERANKERUNGSEISEN
U - SHAPED BAR	« HAARNADEL »
L - SHAPED BAR	WINKELZULAGE
STARTER BAR	ANSCHLUSSEISEN
SPLICE BAR	
ERECTION BAR	MONTAGE EISEN
REINFORCEMENT FOR STRESSES DURING ERECTION WORK	MONTAGEBEWEHRUNG

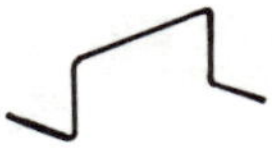

STAY
STEHBÜGEL
MONTAGEBOCK

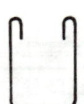

SUPPORT
STEHBÜGEL
OFFENER BÜGEL
FÜR RIPPEN

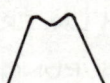

CHAIR
STEHBÜGEL
FÜR BS+G

STIRRUP
BÜGEL

BINDER
BÜGEL
IN STÜTZEN

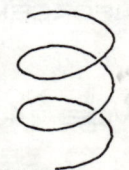

SPIRAL REINFORCEMENT
SPIRALBEWEHRUNG

BENDING AND FIXING OF REINFORCEMENT
BIEGEN UND VERLEGEN

REINFORCEMENT
BEWEHRUNG

REINFORCEMENT BARS IN TABULAR FORM
TABELLENEISEN

BENDING SCHEDULE				STAHLLISTE, BIEGELISTE	
LOCATION BAUTEIL	MARK POS. NR.	No. REQU. ANZAHL	DIA ø	TOTAL LENGTH GESAMTLÄNGE	BENDING FORM
E.G. EDGE BEAM	⑮	25	12	3,150 m	⌐___⌐

ALL BENDING DIMENSIONS ARE OUTER DIMENSIONS
ALLE STAHLMASSE SIND ÄUSSERE MASSE

ALL BAR INTERSECTIONS SHALL BE SECURELY TIED WITH 1,6 mm (16 gauge) ANNEALED SOFT WIRE
ALLE STAHLÜBERKREUZUNGEN SIND MIT AUSGEGLÜHTEM WEICH-DRAHT ø 1,6 FESTZUBINDEN

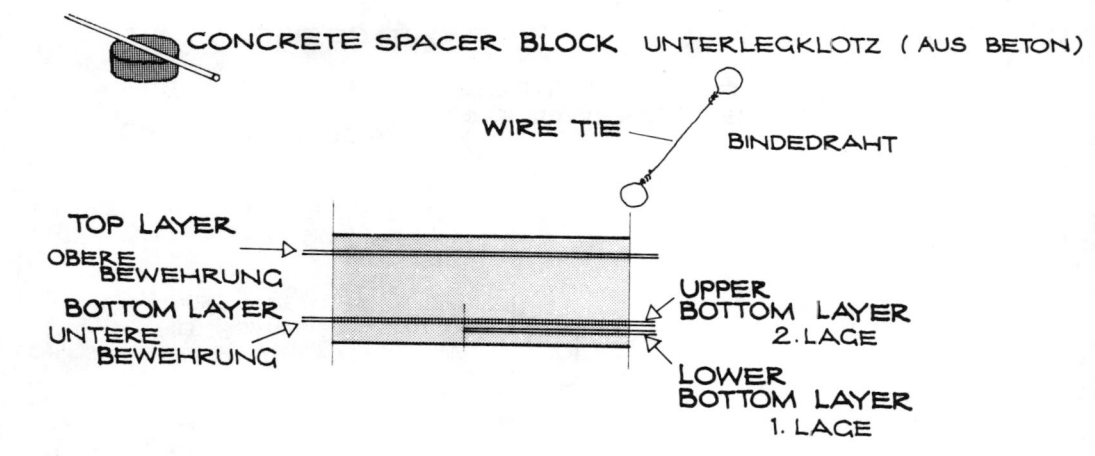

CONCRETE SPACER BLOCK   UNTERLEGKLOTZ (AUS BETON)

WIRE TIE — BINDEDRAHT

TOP LAYER
OBERE BEWEHRUNG
BOTTOM LAYER
UNTERE BEWEHRUNG

UPPER BOTTOM LAYER
2. LAGE
LOWER BOTTOM LAYER
1. LAGE

REINFORCEMENT IN DOUBLE LAYERS
BEWEHRUNG IN ZWEI LAGEN

PROVIDE 40 ø LAPS   ÜBERDECKUNGSSTOSS 40 d

PROVIDE 3 MESHES OVERLAP
           ÜBERDECKUNG BEI BStG 3 MASCHEN

SPECIAL BENDING SPECIFICATIONS FOR RADII DIFFERING FROM STANDARD SPECIFICATIONS
BESONDERE BIEGEANGABEN BEI VON DER NORM ABWEICHENDEN BIEGERADIEN

REINFORCEMENT NOTES

REINFORCEMENT BEWEHRUNG

⊕ CC'S = ⊕
BETWEEN CENTRES / ACHS-ABSTAND

REFER TO WORKING DRAWINGS
    WERKPLÄNE SIND ZU BEACHTEN

6 × 12 mm ⌀ BARS AT 150 mm CC'S TWO-WAYS AT BOTTOM
    6 ⌀ 12, t = 15 cm, KREUZWEISE UNTEN

6 × 12 mm ⌀ MAIN BARS AT 150 mm CC'S ONE-WAY AT BOTTOM
    6 ⌀ 12 LÄNGSEISEN, t = 15 cm, UNTEN

10 mm ⌀ STAYS AT SPACES OF 3,000 m [OR: 3/8" ⌀ STAYS @ 10'0"]
    STEHBÜGEL ⌀ 12, a = 3,0 m VERLEGEN

1 CHAIR PER SQUARE METER
    1 STEHBÜGEL (BStG) PRO m²

10 mm ⌀ STIRRUPS @ 220 mm CC'S THROUGHOUT
    ⌀ 10 BÜGEL t = 22 cm DURCHGEHEND

6 mm ⌀ BINDERS @ (= AT) 300 mm CC'S
    ⌀ 6 BÜGEL (IN STÜTZEN) t = 30 cm

12 mm ⌀ SPACERS   ⌀ 12 VERTEILEREISEN

SPLICE BARS FOR COLUMNS AT CONSTRUCTION JOINTS
    STÜTZENANSCHLUSSBEWEHRUNG AN DEN BETONIERFUGEN

MARK 2 AND MARK 9 BARS ARE TO BE SLUED AROUND (TO STAGGER LAPS)
    POS 2 MIT POS 9 VERSCHWENKEN (ZUR STOSSVERSETZUNG)

BEND TO INSIDE TO AVOID REINFORCEMENT OF COLUMNS
    NACH INNEN VERKRÖPFEN WEGEN DURCHFÜHRUNG DER STÜTZENBEWEHRUNG

BAR TO BE BENT ASIDE   STAB IN ANDERE EBENE VERKRÖPFEN

DRILL HOLES ⌀ 25 mm AT 150 mm CC'S ON THE SITE THRO' WEB OF M.S. PROFILE TO INTERLACE REINFORCEMENT OF SLAB
    DER STEG DES WALZPROFILTRÄGERS IST ZUR DURCHFÜHRUNG DER DECKENBEWEHRUNG BAUSEITS ZU DURCHBOHREN
    ⌀ 25 ALLE 15 cm

FORMWORK
SCHALARBEITEN

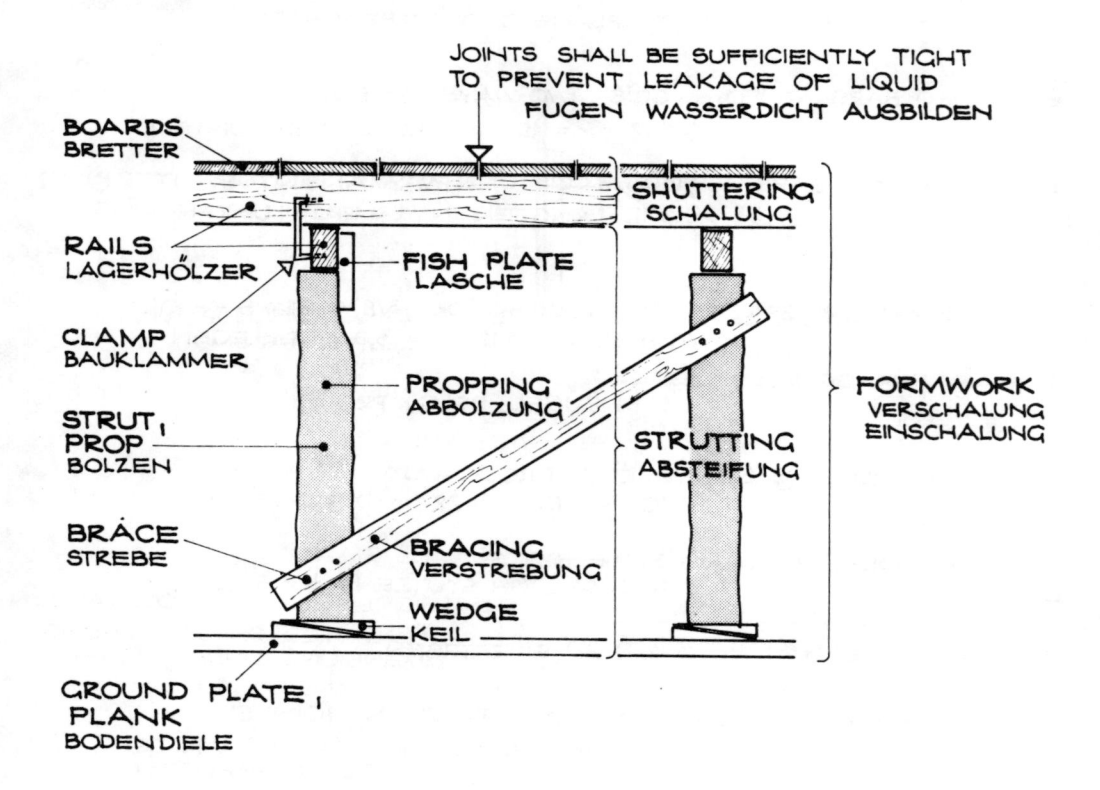

JOINTS SHALL BE SUFFICIENTLY TIGHT TO PREVENT LEAKAGE OF LIQUID
FUGEN WASSERDICHT AUSBILDEN

- BOARDS / BRETTER
- RAILS / LAGERHÖLZER
- FISH PLATE / LASCHE
- CLAMP / BAUKLAMMER
- STRUT, PROP / BOLZEN
- PROPPING / ABBOLZUNG
- SHUTTERING / SCHALUNG
- STRUTTING / ABSTEIFUNG
- BRACE / STREBE
- BRACING / VERSTREBUNG
- WEDGE / KEIL
- GROUND PLATE, PLANK / BODENDIELE
- FORMWORK / VERSCHALUNG, EINSCHALUNG

SHUTTERING
SHUTTERING SHALL IMPART A SMOOTH FINISH TO THE RESULTANT CONCRETE SURFACE

DIE SCHALUNG MUSS (SOLL) DEM BETON EINE GLATTE OBERFLÄCHE VERLEIHEN

STRUTTING
VERTICAL STRUTTING SHALL BE SUFFICIENTLY STRONG TO AFFORD THE REQUIRED SUPPORT WITHOUT SETTLEMENT

DIE VERTIKALE ABSTEIFUNG MUSS GENÜGEND TRAGFÄHIG SEIN (DEN ANFALLENDEN LASTEN) DEN ERFORDERLICHEN ABSTÜTZWIDERSTAND OHNE SETZUNGEN ZU GEWÄHREN

STRIKING — AUSSCHALEN
EASING, STRIKING AND REMOVING OF FORMWORK SHALL BE DONE WITHOUT SHOCK OR VIBRATION

DAS LÖSEN, ABBAUEN UND ENTFERNEN DER SCHALUNG MUSS STOSS- UND ERSCHÜTTERUNGSFREI AUSGEFÜHRT WERDEN

STRIKING TIMES   AUSSCHALZEITEN

DETAILS

FORMWORK
SCHALARBEITEN

FORMWORK MAY BE IN STEEL OR IN TIMBER
SCHALUNG KANN IN STAHL ODER HOLZ AUSGEFÜHRT WERDEN

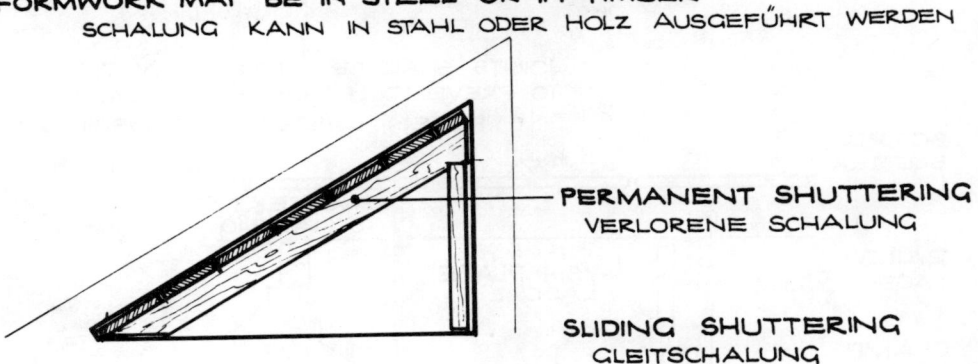

PERMANENT SHUTTERING
VERLORENE SCHALUNG

SLIDING SHUTTERING
GLEITSCHALUNG

SHUTTER   LADEN
E.G. ROLLER SHUTTER   Z.B. ROLLADEN
WINDOW SHUTTER   FENSTERLADEN

SHUTTERBOARD   SCHALTAFEL (AUS HOLZ)
STEEL SHUTTER   SCHALBLECH (AUS STAHL)

TOLERANCES   UNGENAUIGKEITEN, MASSTOLERANZEN

A TOLERANCE OF UP TO 6 mm WILL BE PERMITTED ON DIMENSIONS OF 3,0 m AND OVER AND 3 mm ON DIMENSIONS UNDER 3,0 m.
UNGENAUIGKEITEN BIS ZU 6 mm WERDEN BEI MASSEN $\geq 3,0$ m ZUGELASSEN, BIS ZU 3 mm BEI MASSEN $< 3,0$ m

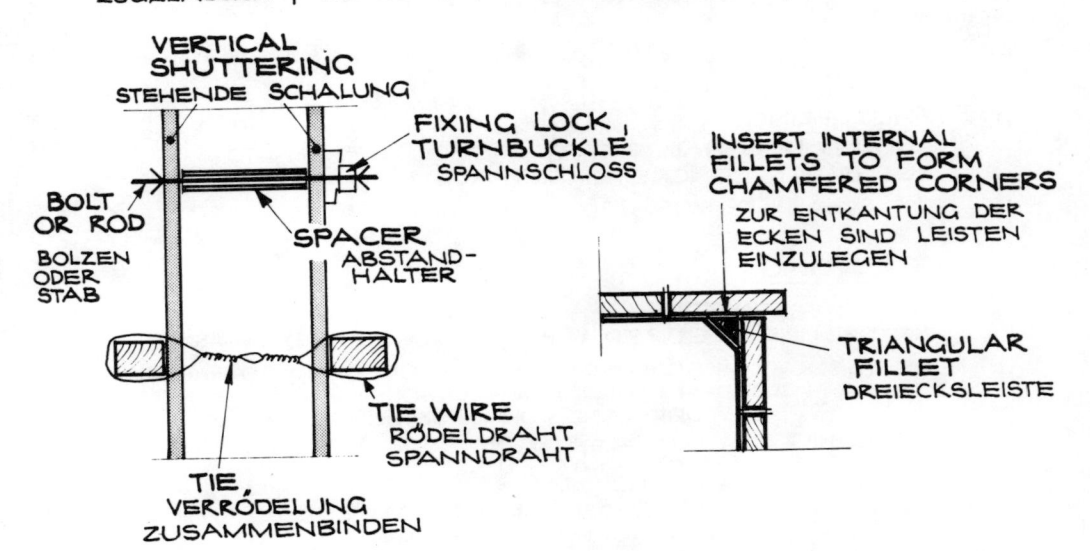

VERTICAL SHUTTERING
STEHENDE SCHALUNG

FIXING LOCK, TURNBUCKLE
SPANNSCHLOSS

BOLT OR ROD
BOLZEN ODER STAB

SPACER
ABSTANDHALTER

TIE WIRE
RÖDELDRAHT
SPANNDRAHT

TIE,
VERRÖDELUNG
ZUSAMMENBINDEN

INSERT INTERNAL FILLETS TO FORM CHAMFERED CORNERS
ZUR ENTKANTUNG DER ECKEN SIND LEISTEN EINZULEGEN

TRIANGULAR FILLET
DREIECKSLEISTE

FOR FORMWORK TOOLS SEE CARPENTRY TOOLS
SCHALWERKZEUGE SIEHE UNTER ZIMMERMANNS WERKZEUGE

BRICKWORK, MASONRY
MAURERARBEITEN

PLASTER WORK
VERPUTZ

SETT PAVING
PFLASTER

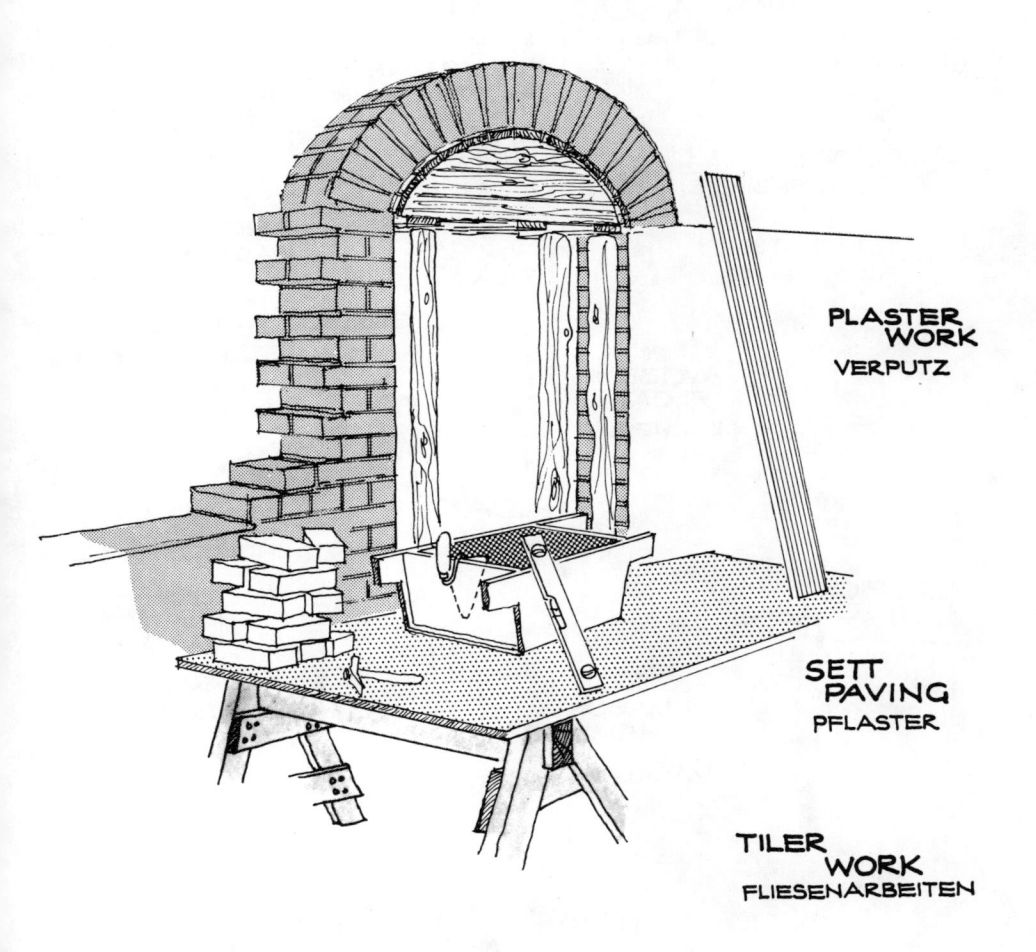

TILER WORK
FLIESENARBEITEN

MAURERARBEITEN   # BRICKLAYER'S TOOLS   BRICKWORK
MAURERWERKZEUGE

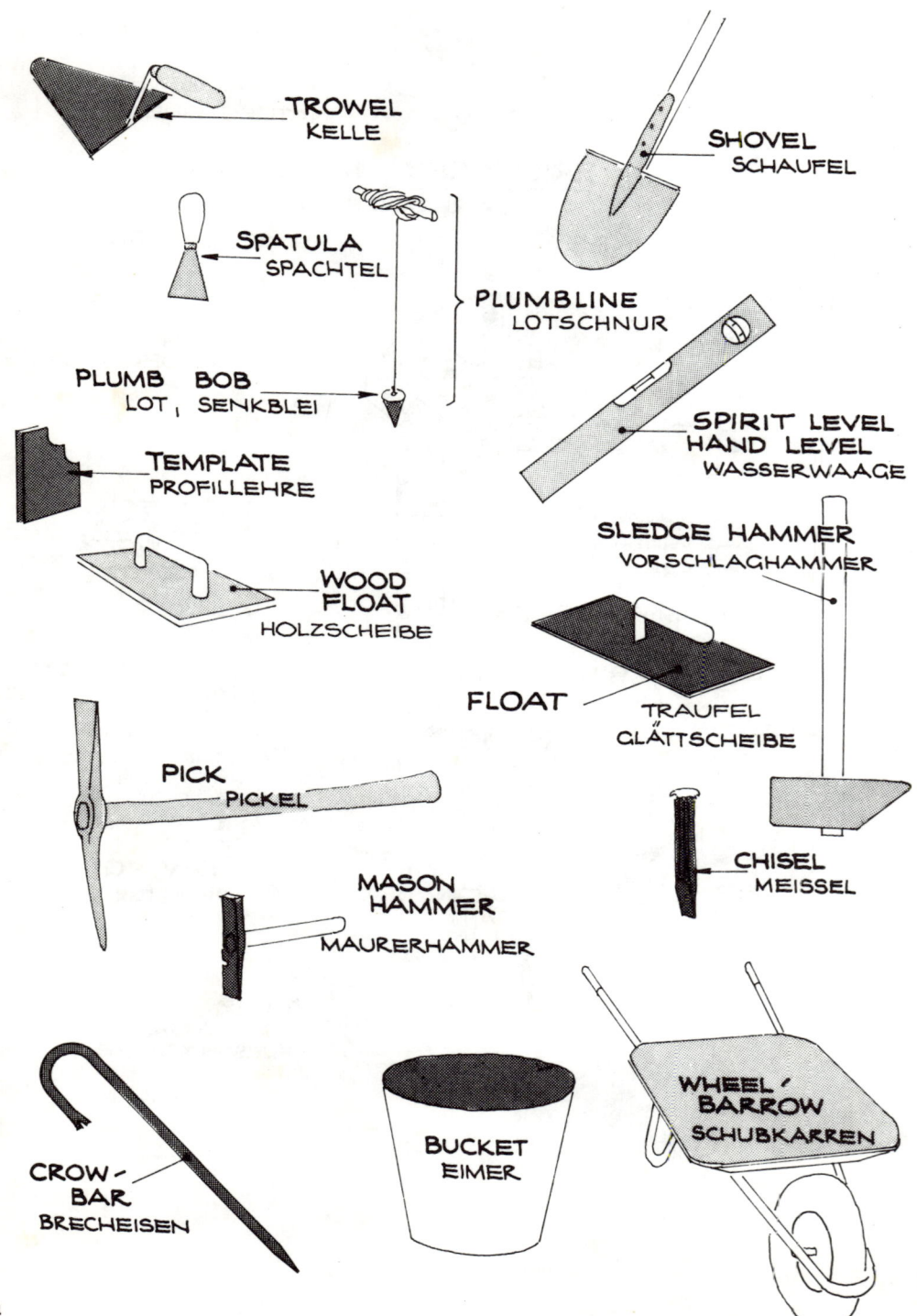

- TROWEL — KELLE
- SHOVEL — SCHAUFEL
- SPATULA — SPACHTEL
- PLUMBLINE — LOTSCHNUR
- PLUMB BOB — LOT, SENKBLEI
- SPIRIT LEVEL / HAND LEVEL — WASSERWAAGE
- TEMPLATE — PROFILLEHRE
- SLEDGE HAMMER — VORSCHLAGHAMMER
- WOOD FLOAT — HOLZSCHEIBE
- FLOAT — TRAUFEL, GLÄTTSCHEIBE
- PICK — PICKEL
- CHISEL — MEISSEL
- MASON HAMMER — MAURERHAMMER
- CROWBAR — BRECHEISEN
- BUCKET — EIMER
- WHEELBARROW — SCHUBKARREN

MAURERARBEITEN   **STONES AND BRICKS**   BRICKWORK
STEINE UND ZIEGEL

NATURAL STONES — NATÜRLICHE STEINE / NATURSTEINE

MAN-MADE STONES — KÜNSTLICHE STEINE

BRICKS:
   STOCK BRICK — GEWÖHNLICHER MAUERZIEGEL

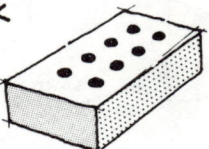

   PERFORATED BRICK — LOCHZIEGEL
     VERTICALLY PERFORATED BRICK — HOCHLOCHZIEGEL
     HORIZONTALLY PERFOR. BRICK — LANGLOCHZIEGEL

   CLINKER BRICK — KLINKER

   FACE BRICK } VERBLENDSTEIN
   FACING BRICK } SICHTMAUERSTEIN
     SIDE FACED BRICK — LÄNGSSICHTSTEIN
     SIDE AND ONE END FACED BRICK — ECKSICHTSTEIN

NOTE: BRICKS ARE TO BE HARD, WELL BURNT, REASONABLY UNIFORM IN SIZE AND SHAPE AND EQUAL TO SAMPLE

ZIEGEL MÜSSEN HART SEIN, GUT GEBRANNT, (ZIEMLICH) EINHEITLICH IM FORMAT UND GLEICH IM AUSEHEN

BLOCKS:
   CONCRETE BLOCK — BETONBLOCK

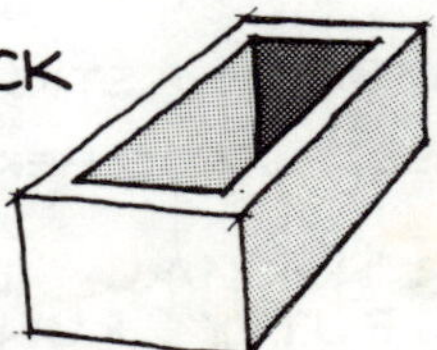

   HOLLOW BLOCK OR HOLLOW CONCRETE BLOCK — HOHLBLOCK

   AERATED CONCRETE BLOCK — GASBETONBLOCK

   LIGHT-WEIGHT BUILDING SLAB — LEICHTBAUPLATTE

MAURERARBEITEN   **MORTAR**   BRICKWORK
MÖRTEL

CEMENT MORTAR   ZEMENTMÖRTEL

SHALL BE COMPOSED OF   SOLL AUS 6 RAUMTEILEN
6 PARTS BY VOLUME OF SAND TO   SAND UND 1 RAUMTEIL (RT)
1 PART BY VOLUME OF CEMENT   ZEMENT GEMISCHT WERDEN
(OR OTHER PROPORTIONS)   (ODER ANDERE MISCHUNGEN)

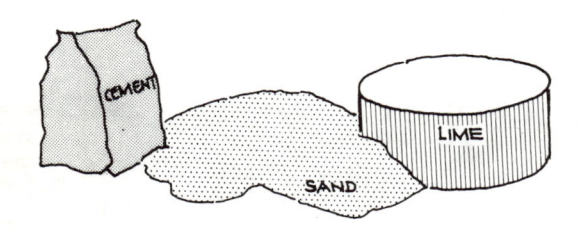

COMPO MORTAR, GAUGED MORTAR   KALKZEMENTMÖRTEL

SHALL BE COMPOSED OF   SOLL AUS 9 RAUMTEILEN
9 PARTS SAND,   SAND, 2 RT KALK UND 1 RT
2 PARTS LIME AND   ZEMENT GEMISCHT WERDEN
1 PART CEMENT   (ODER ANDERE MISCHUNGEN)
(OR OTHER PROPORTIONS)

LIME MORTAR   KALKMÖRTEL

- USED FOR PLASTER ONLY -
- NUR ZUM VERPUTZEN -

SHALL BE COMPOSED OF
5 PARTS SAND TO
1 PART LIME
(OR OTHER PROPORTIONS)

NOTE:

     LIME SHALL BE PLASTIC HYDRATED LIME (IN BAGS)...
       KALK MUSS PLASTISCHER HYDRAULISCHER KALK SEIN
                                        (SACKKALK)

     ..... OR NORMAL LIME GAINED FROM SLAKED
WHITE LIME PUTTY RUN THROUGH A FINE SIEVE
INTO A RECEPTACLE WHERE IT IS TO REMAIN
UNTIL IT ATTAINS THE PROPER CONSISTENCY FOR USE

     .... ODER NORMALER LUFTKALK, DER GEWONNEN WURDE
AUS NASS GELÖSCHTEM KALKBREI, FEIN GESIEBT UND IN
EINEM BEHÄLTER BIS ZUR ERREICHUNG DER RICHTIGEN
DICKE (KONSISTENZ) GELAGERT

MAURERARBEITEN   **JOINTS AND BONDS**   BRICKWORK
FUGEN UND VERBÄNDE

JOINTS:
FUGEN

JOINT, PERPEND OR TRANSVERSAL JOINT
STOSSFUGE

MORTAR BED OR LONGITUDINAL JOINT
LAGER FUGE

STAGGERED JOINT
VERSETZTE FUGE

POINTED JOINT
SICHTFUGE

RAKED JOINT
AUSGEKRATZTE FUGE

HALFROUND POINTING
HOHLFUGE

SQUARE RECESSED POINTING
ECKIG ZURÜCKGESETZTE FUGE

COURSES OR LAYERS:
SCHICHTEN

FLAT COURSE
STRETCHER COURSE
LÄUFERSCHICHT

BRICK ON EDGE COURSE
HOCHKANT-SCHICHT

SOLDIER COURSE
STEHENDE SCHICHT

BONDS:
VERBÄNDE

STRETCHER BOND — LÄUFERVERBAND

HEADER BOND — BINDERVERBAND

ENGLISH BOND

ALLE KOMBINATIONEN VON LÄUFER-BINDER-VERBÄNDEN WIE BLOCK-, KREUZ-, GOTIKVERBAND ETC WERDEN ALLGEMEIN ALS «ENGLISH BOND» BEZEICHNET

ALL COMBINATIONS OF "LÄUFER-BINDER-VERBÄNDEN" LIKE "BLOCK-, KREUZ-, GOTIKVERBAND" ETC. WILL BE GENERALY TERMED AS "ENGLISH BOND" (IN BRITAIN)

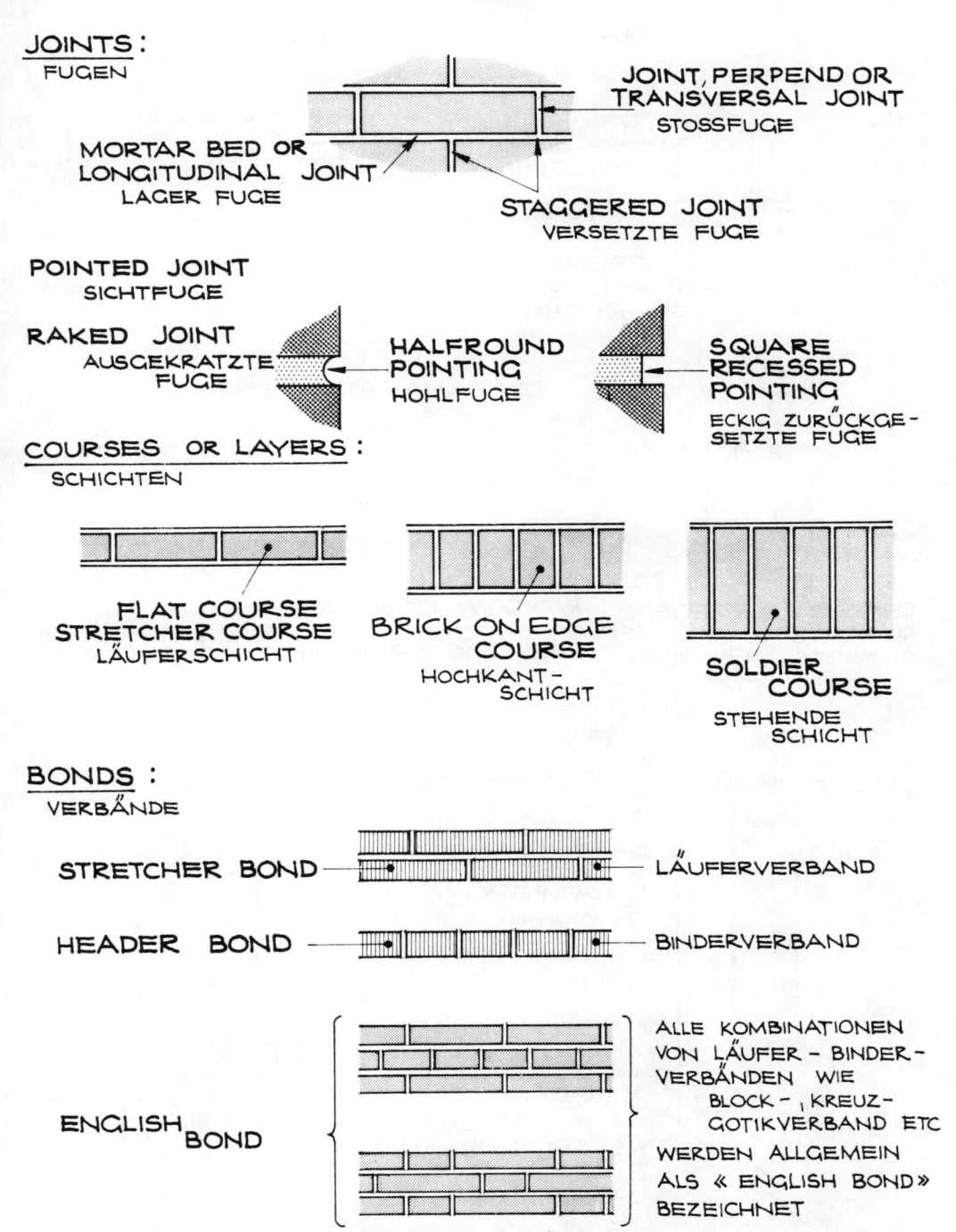

MAURERARBEITEN **BRICK STRUCTURES** BRICKWORK
GEMAUERTE BAUTEILE

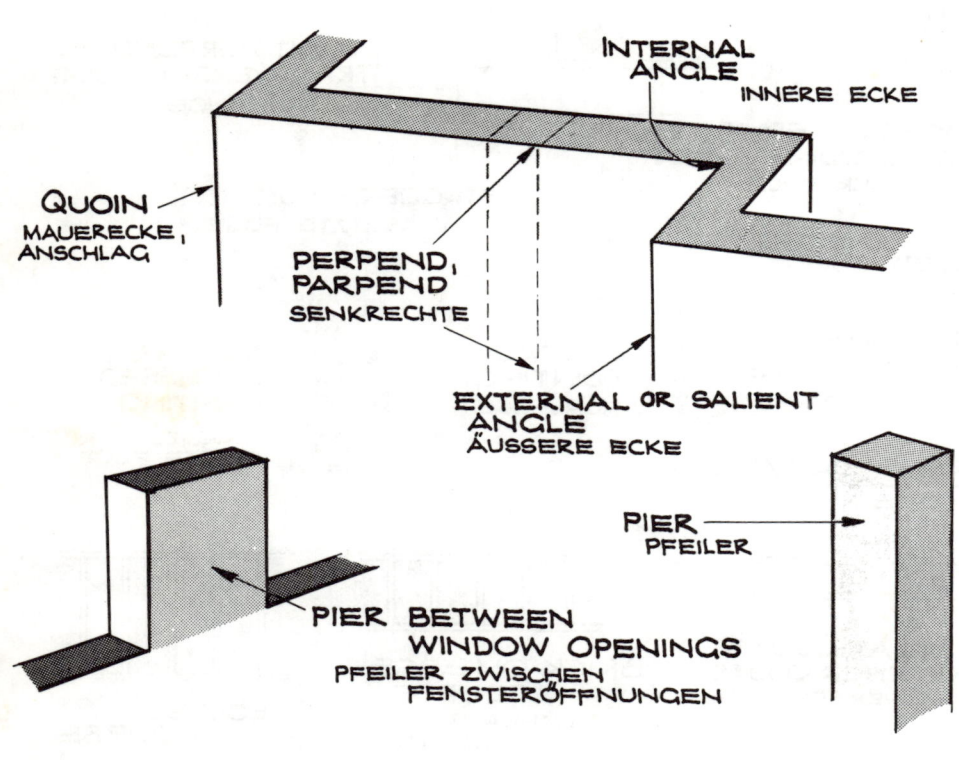

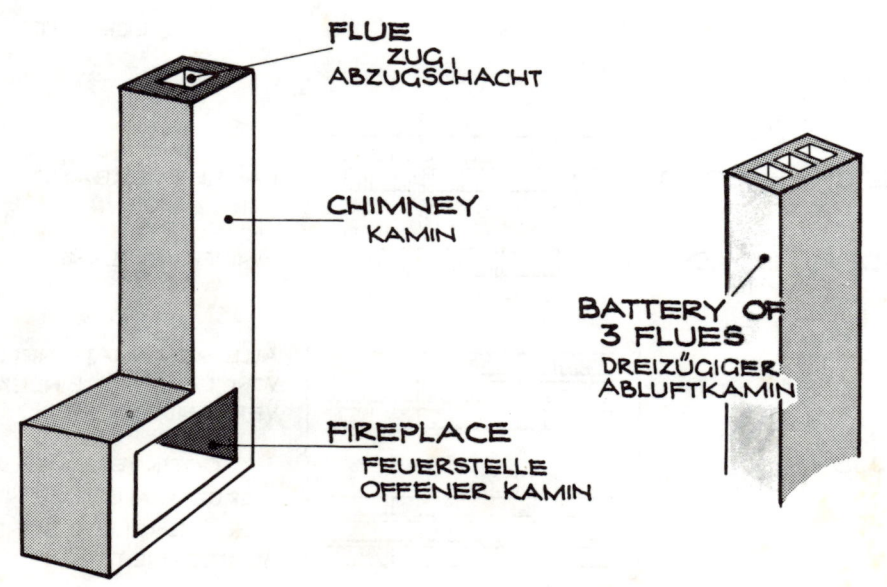

INTERNAL ANGLE — INNERE ECKE

QUOIN — MAUERECKE, ANSCHLAG

PERPEND, PARPEND — SENKRECHTE

EXTERNAL OR SALIENT ANGLE — ÄUSSERE ECKE

PIER — PFEILER

PIER BETWEEN WINDOW OPENINGS — PFEILER ZWISCHEN FENSTERÖFFNUNGEN

FLUE — ZUG, ABZUGSCHACHT

CHIMNEY — KAMIN

BATTERY OF 3 FLUES — DREIZÜGIGER ABLUFTKAMIN

FIREPLACE — FEUERSTELLE, OFFENER KAMIN

MAURERARBEITEN **BRICK STRUCTURES** BRICKWORK
Gemauerte Bauteile

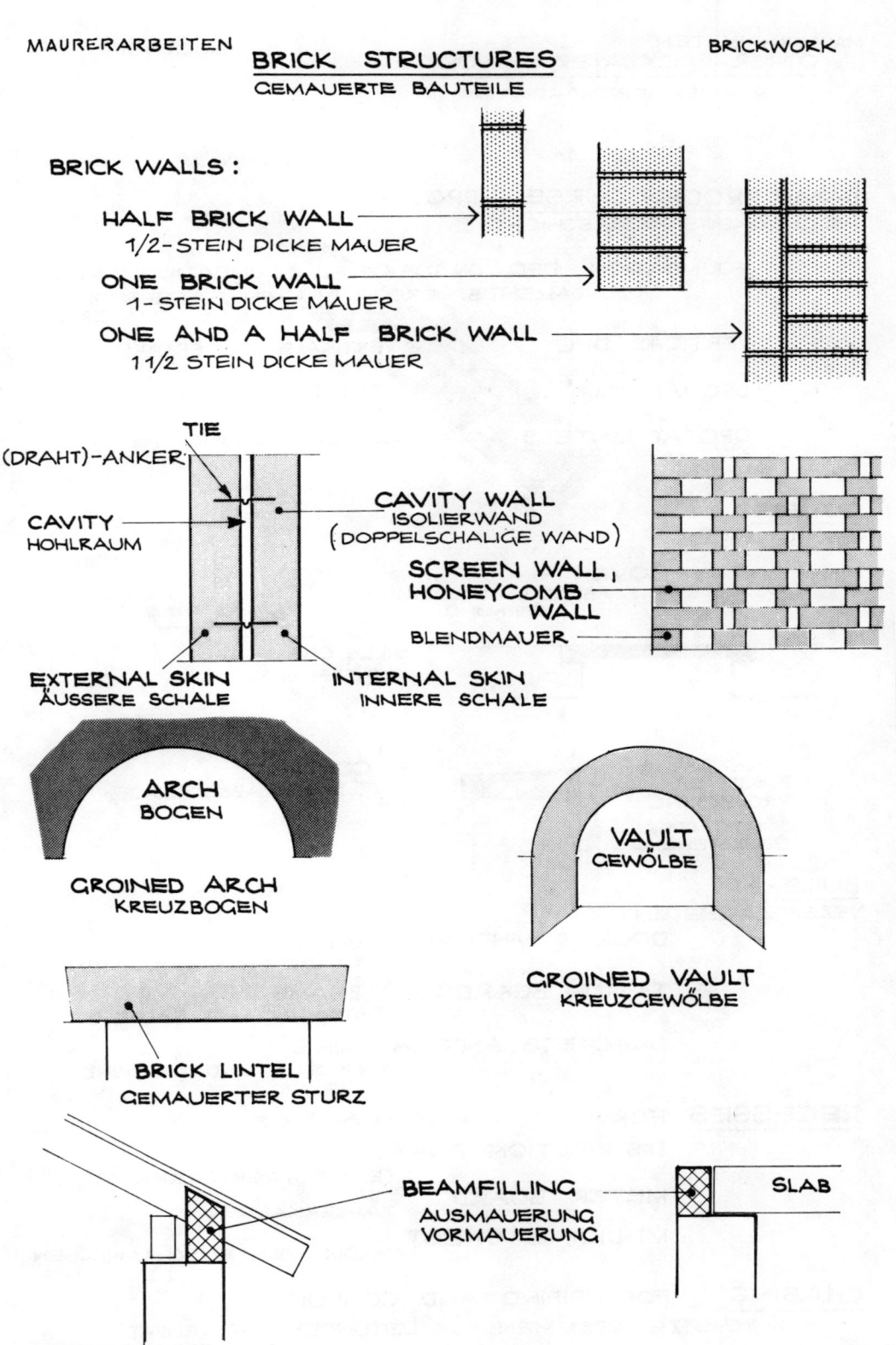

BRICK WALLS:

HALF BRICK WALL
1/2 - STEIN DICKE MAUER

ONE BRICK WALL
1 - STEIN DICKE MAUER

ONE AND A HALF BRICK WALL
1 1/2 STEIN DICKE MAUER

TIE
(DRAHT)-ANKER

CAVITY
HOHLRAUM

CAVITY WALL
ISOLIERWAND
(DOPPELSCHALIGE WAND)

SCREEN WALL,
HONEYCOMB
WALL
BLENDMAUER

EXTERNAL SKIN
ÄUSSERE SCHALE

INTERNAL SKIN
INNERE SCHALE

ARCH
BOGEN

GROINED ARCH
KREUZBOGEN

VAULT
GEWÖLBE

GROINED VAULT
KREUZGEWÖLBE

BRICK LINTEL
GEMAUERTER STURZ

BEAMFILLING
AUSMAUERUNG
VORMAUERUNG

SLAB

MAURERARBEITEN
OTHER BRICKLAYER'S WORK   BRICKWORK
MAURERNEBENARBEITEN

DAMP PROOF COURSE (DPC):
FEUCHTIGKEITSISOLIERSCHICHT

     FOUNDATION DPC ON TOP OF ALL FOUNDATION WALLS
         FUNDAMENTISOLIERUNG IN ALLEN AUFGEHENDEN MAUERN

     VERTICAL DPC   VERTIKALE ISOLIERUNG

     DPC AT SILL   FENSTERBANKISOLIERUNG

     DPC AT LINTELS   STURZISOLIERUNG

150 mm STÖSSE

COVERS:
ABDECKUNGEN

COVER — ABDECKUNG EINER ÖFFNUNG

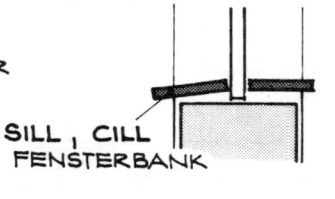

SILL, CILL — FENSTERBANK

FLUSH OR PROJECTING
BÜNDIG ODER ÜBERSTEHEND

COPING — MAUERABDECKUNG

BUILD-IN:
VERSETZARBEITEN

     DOOR & WINDOW FRAMES
         TÜR- UND FENSTERRAHMEN

     T-IRON BEARERS & BEAMS
         T-TRÄGER UND BALKEN

     BRACKETS AND THE LIKE
         KONSOLEN UND ÄHNLICHES

RECESSES FOR   AUSSPARUNGEN FÜR
     DISTRIBUTION BOARD
         (ELEKTR.) VERTEILERKASTEN

     METER BOARD
         ZÄHLERKASTEN

     MEDICINE CHEST
         MEDIKAMENTENSCHRÄNKCHEN

CHASING   FOR PIPING AND CONDUITS
     SCHLITZE STEMMEN FÜR LEITUNGEN UND (ELEKT.) SCHUTZROHRE

BRICK LAYING NOTES
MAUERRICHTLINIEN

BRICKWORK
MAURERARBEITEN

STOCKBRICK WORK

ALL BRICKWORK IS TO BE BUILT IN ENGLISH BOND EXCEPT HALF BRICK AND CAVITY WALLS WHICH ARE TO BE IN STRETCHER BOND. NO FALSE HEADERS AND NONE BUT WHOLE BRICKS EXCEPT WHERE LEGITIMATELY REQUIRED TO FORM BOND ARE TO BE USED. ALL PERPENDS AND ANGLES ARE TO BE PLUMB. (= VERTICAL) MORTAR BEDS ARE NOT TO EXCEED 10 mm THICKNESS AND JOINTS ARE TO BE FLUSHED UP SOLID AT EVERY COURSE. WHERE WALLS ARE TO BE PLASTERED JOINTS TO BE WELL RAKED OUT TO FORM KEY FOR PLASTER.

MAUERWERK (STANDARDAUSFÜHRUNG)
DAS GESAMTE MAUERWERK IST IM „ENGLISCHEN VERBAND" AUSZUFÜHREN, AUSSER HALBZIEGEL- UND DOPPELSCHALIGEN MAUERN, DIE IM LÄUFERVERBAND GEMAUERT WERDEN. KEINE „FALSCHEN BINDER" UND NUR GANZE ZIEGELSTEINE VERWENDEN, AUSSER SIE WERDEN AUSDRÜCKLICH ZUR HERSTELLUNG EINES VERBANDES GEBRAUCHT. ALLE WANDSTÄRKEN STREKKER UND WINKEL SIND INS LOT ZU BRINGEN (=SENKRECHT). DAS MÖRTELBETT SOLL 10MM DICKE NICHT ÜBERSCHREITEN UND DIE FUGEN SIND SORGFÄLTIG IN JEDER LAGE AUSZUFÜLLEN. SOLL DIE MAUER VERPUTZT WERDEN, SIND DIE FUGEN AUFZURAUHEN, UM EINEN GUTEN HAFTGRUND FÜR PUTZ ABZUGEBEN.

FACEBRICK WORK

ALL FACEWORK IS TO BE BUILT TO A FAIR FACE AND POINTED WITH A SQUARE 10 mm RECESSED JOINT AS WORK PROCEEDS, AND ALL SUCH FACES ARE TO BE PROTECTED FROM INJURY. ONE BRICK WALLS THAT ARE SPECIFIED AS FACED ARE TO BE BUILT IN TWO HALF BRICK SKINS LAID IN STRETCHER BOND AND TIED TOGETHER WITH GALVANISED CRIMPED WIRE TIES.

SICHTMAUERWERK
VERBLENDUNGEN SIND ALS SICHTFLÄCHEN HERZUSTELLEN UND WERDEN NACH ARBEITSFORTGANG MIT 10MM TIEFEN QUADRATISCHEN FUGEN VERSEHEN. ALLE OBERFLÄCHEN SIND VOR BESCHÄDIGUNGEN ZU SCHÜTZEN. EINSCHALIGE, ZUR VERKLEIDUNG VORGESEHENE MAUERN KÖNNEN AUCH AUS ZWEI SCHALEN MIT HALBEN STEINEN IM LÄUFERVERBAND HERGESTELLT UND MIT VERZINKTEN UND GEKRÖPFTEN DRAHTANKERN GESICHERT WERDEN.

CAVITY WALLS

CAVITY WALLS ARE TO BE BUILT UP IN TWO HALF BRICK SKINS TIED TOGETHER WITH GALVANISED CRIMP WIRE TIES, CAREFULLY LAID.
CAVITIES MUST BE KEPT FREE OF MORTAR DROPPINGS OR OTHER MATTER BY MOVEABLE BOARDS, AND TEMPORARY OPENINGS MUST BE LEFT AT PLINTH LEVEL THROUGH WHICH THE DROPPINGS CAN BE REMOVED.

ZWEISCHALIGES MAUERWERK (ISOLIERMAUERWERK)
ISOLIERMAUERWERK IST ZWEISCHALIG MIT HALBEN STEINEN HERZUSTELLEN UND MIT VERZINKTEN, GEKRÖPFTEN DRAHTANKERN SORGFÄLTIG ZU SICHERN. DIE HOHLRÄUME MÜSSEN DURCH (FUGENLEHREN ODER) BEWEGBARE LATTEN VON MÖRTELBRÜCKEN UND ANDEREN MATERIALIEN FREIGEHALTEN WERDEN. DURCH HILFSÖFFNUNGEN AUF SOCKELHÖHE KÖNNEN MÖRTELRESTE ENTFERNT WERDEN.

MAURERARBEITEN   **PLASTER WORK**   BRICKWORK
VERPUTZARBEITEN

GROUT,
SLURRY, SLUDGE   VORSPRITZWURF
(TO GIVE KEY FOR PLASTER)   (ALS PUTZUNTERLAGE)

PLASTER TROWEL
PUTZKELLE

RENDERING OR PLASTER   VERPUTZ, PUTZ

E.G. EXTERNAL THREE COAT RENDERING   3-LAGIGER AUSSENPUTZ
E.G. INTERNAL TWO COAT PLASTER   2-LAGIGER INNENPUTZ

CEMENT PLASTER   ZEMENTPUTZ

LIME PLASTER   KALKPUTZ

TWO COAT LIME PLASTER   ZWEILAGIGER KALKPUTZ

BAGGED PLASTER   (MIT SACK ABGERIEBENER)
SPRITZPUTZ

TYROLEAN PLASTER   TIROLER RAUHPUTZ

STUCCO WORK   STUCK-, GESIMSARBEIT,
PLASTER OF PARIS WORK   GIPSARBEIT

OTHER RELATED WORK
ANDERE VERWANDTE ARBEITEN

METAL LATHING   RABITZ-ARBEIT

CORNER PROTECTION STRIP
EDGE PROTECTION STRIP
     KANTENSCHUTZEISEN

V-JOINT
     OFFENE FUGE, SCHWEDENSCHNITT

HOLLOW ROUNDED
FILLET
     HOHLKEHLE

WOOD FLOATED   MIT DER SCHEIBE ABGERIEBEN

STEEL TROWELLED   MIT GLÄTTSCHEIBE GEGLÄTTET

MAURERARBEITEN   **PAVIOR WORK**   BRICKWORK
                     PFLASTERARBEITEN

PAVED FLOORS   PFLASTER, BODENPLATTENBELÄGE

SCREED
ESTRICH

     CEMENT SCREED   ZEMENTESTRICH

     GRANO = GRANOLITHIC SCREED   ZEMENTESTRICH MIT GRANITZUSCHLAG
     (WITH GRANITE CHIPPINGS)

     NON-SLIP GRANOLITHIC   RUTSCHFESTER ZEMENTESTRICH (MIT KARBORUND OBERSCHICHT)
     (WITH CARBORUNDUM TOPPING)

     TINTED GRANOLITHIC   GEFÄRBTER ZEMENTESTRICH

TERRAZZO   KUNSTSTEIN, TERRAZZO

     IN-SITU TERRAZZO   ORTTERRAZZO

     TERRAZZO TILE   TERRAZZOFLIESE

     TERRAZZO SLAB   TERRAZZOPLATTE

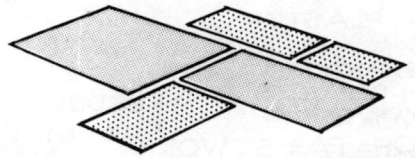

PAVING
PFLASTER

     CEMENT PAVING   ZEMENTPFLASTER

     ACID RESISTANT PAVING   SÄUREFESTES PFLASTER (MIT EPOXY OBERSCHICHT)
     (WITH EPOXY TOPPING)

     BRICK PAVING   ZIEGELPFLASTER

     STONE PAVING   STEINPFLASTER

     WOOD BLOCK PAVING   HOLZPFLASTER

SKIRTINGS   SOCKELLEISTEN

     COVED SKIRTING, HOLLOW ROUNDED SKIRTING   SOCKEL MIT HOHLKEHLE

     WITH V-JOINT TO PLASTER   MIT OFFENER FUGE ZUM VERPUTZ

MAURERARBEITEN — BRICKWORK
TILING
FLIESENARBEITEN

WALL TILES: WANDFLIESEN

GLAZED TILE =   GLASIERTE PORZELLANFLIESE
GLAZED PORCELAIN TILE

CERAMIC TILE   KERAMISCHE FLIESE

FIXED TO PLASTER BY SPECIAL ADHESIVE   SPACER LUG — FUGENKLOTZ
MIT FLIESENKLEBER AUF PUTZ GEKLEBT

FLOOR TILES AND COPINGS: BODENFLIESEN UND ABDECKUNGEN

QUARRY TILE   BODENPLATTE (AUS STEINBRUCHABFÄLLEN)
TERRAZZO TILE   TERRAZZOFLIESE
SLATE   SCHIEFERPLATTE

EMBEDDED IN CEMENT MORTAR   REEDED TILE FOR STAIRS
IM MÖRTELBETT   GERIFFELTE (TRITTSICHERE) FLIESE

JOINTS: FUGEN

POINTED IN WHITE CEMENT
MIT WEISSZEMENT AUSGEFUGT

OTHER:

RECESSED TOILET PAPER HOLDER
EINGELASSENER PAPIERHALTER
RECESSED SOAP DISH   SEIFENSCHALE
TOWEL HOOK   HANDTUCHHALTER
BATHROOM CHEST   SPIEGELSCHRANK, TOILETTENSCHRANK

MAURERARBEITEN   **PREFABRICATED BUILDING**   BRICKWORK
FERTIGTEIL-BAUWEISE

PANELS:
- WALL PANEL — WANDPLATTE
- FLOOR PANEL — BODENPLATTE
- ROOF PANEL — DACHPLATTE

OF CONCRETE — AUS BETON
  AERATED CONCRETE — GASBETON
  SANDWICH BOARD — MEHRSCHALIGE ISOLIERPLATTE

TO ASSEMBLE — ZUSAMMENBAUEN
TO ERECT — AUFSTELLEN, MONTIEREN

WEDGING — VERKEILUNG
SLEEVE FOUNDATION — KÖCHER-FUNDAMENT
CENTERING PIN — ZENTRIERSTIFT

CAST-IN FIXING RAILS — EINBETONIERTE BEFESTIGUNGSSCHIENEN

CURTAIN WALL — VORGEHÄNGTE FASSADE

FIXING WITH — BEFESTIGUNG MIT
  RAWL PLUG — RAULDÜBEL

PLUG FIXED WITH FILLER/ADHESIVE (E.G. TWO-PACK EPOXY RESIN)
DÜBEL MIT KLEBER (MIT ZWEIKOMPONENTENKLEBER) BEFESTIGT

STEEL- AND METALWORK
STAHL - UND METALLARBEITEN

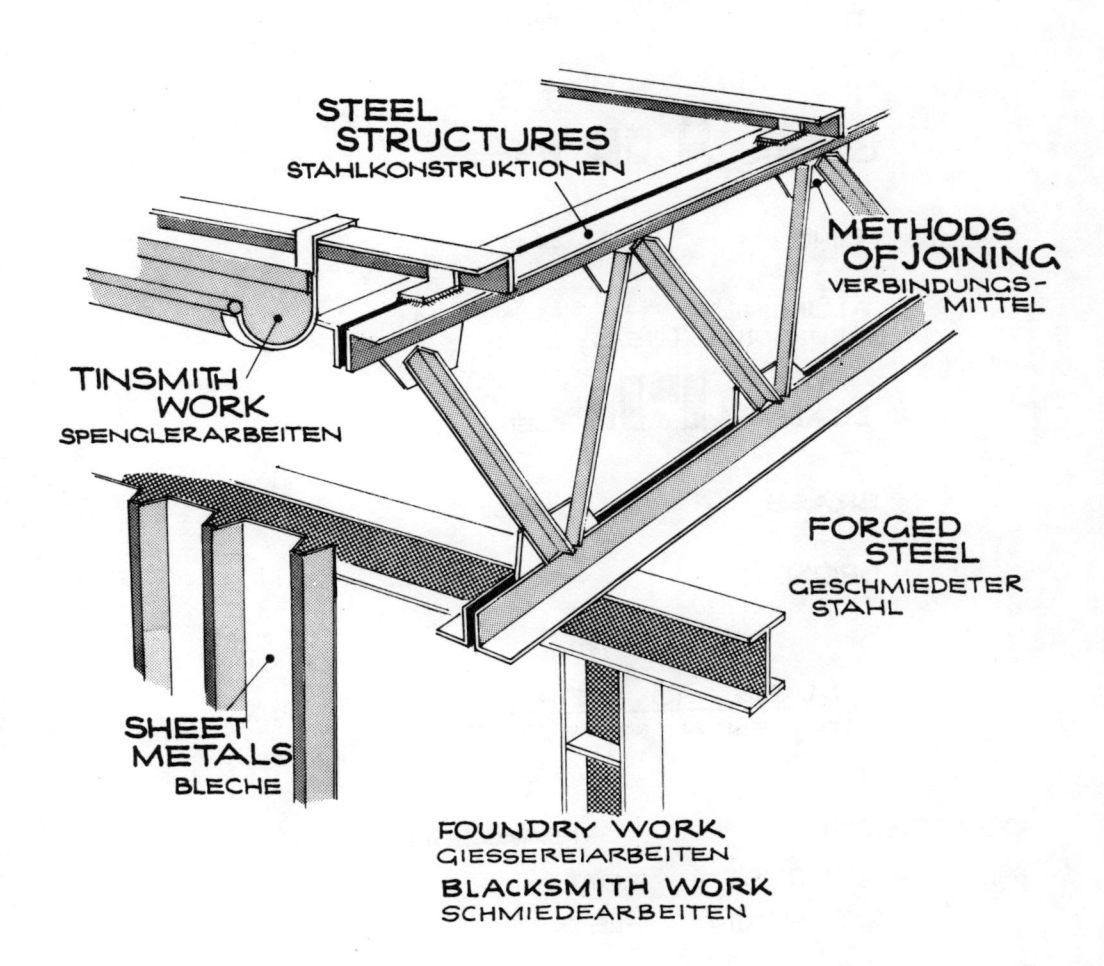

STEEL STRUCTURES
STAHLKONSTRUKTIONEN

METHODS OF JOINING
VERBINDUNGSMITTEL

TINSMITH WORK
SPENGLERARBEITEN

FORGED STEEL
GESCHMIEDETER STAHL

SHEET METALS
BLECHE

FOUNDRY WORK
GIESSEREIARBEITEN

BLACKSMITH WORK
SCHMIEDEARBEITEN

METALS
METALLE

STEELWORK
STAHL-UND METALLARBEITEN

STEEL	**Fe**	STAHL
MILD STEEL		
FLUSS-STAHL		
IRON		EISEN
TIN		ZINN
COPPER	**Cu**	KUPFER
ZINC		ZINK
ALUMINIUM		ALUMINIUM
ALUMINUM (USA)		
LEAD	**Pb**	BLEI
BRASS		MESSING
BRONZE		BRONZE

DAMAGE TO METAL PARTS
SCHÄDEN AN METALLTEILEN

CORROSION	KORROSION
RUST	ROST
STAIN	FLECKEN
DEFORMATION	VERFORMUNG
DEFLECTION	DURCHBIEGUNG

STEEL WORKING TOOLS
WERKZEUG ZUR STAHLBEARBEITUNG

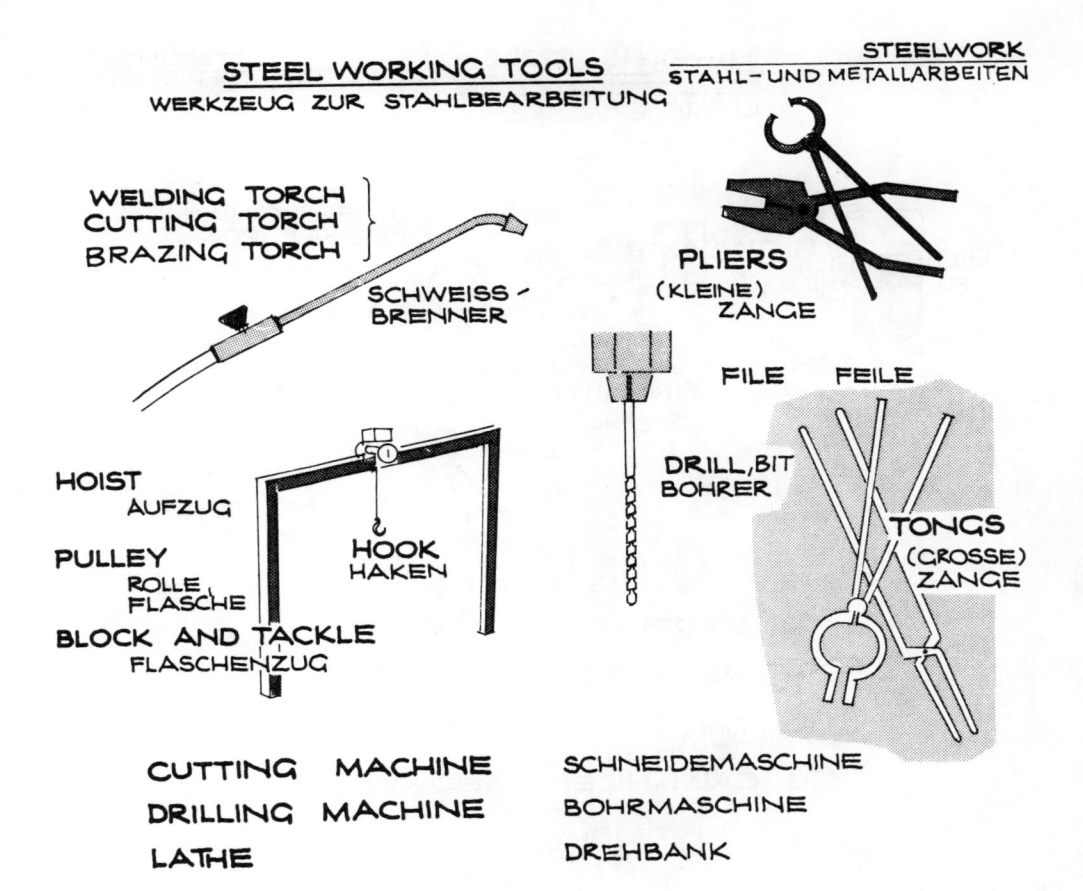

STEELWORK
STAHL- UND METALLARBEITEN

WELDING TORCH
CUTTING TORCH
BRAZING TORCH
SCHWEISS-BRENNER

PLIERS (KLEINE) ZANGE

FILE — FEILE

DRILL, BIT — BOHRER

TONGS (GROSSE) ZANGE

HOIST — AUFZUG
PULLEY — ROLLE, FLASCHE
BLOCK AND TACKLE — FLASCHENZUG
HOOK — HAKEN

CUTTING MACHINE — SCHNEIDEMASCHINE
DRILLING MACHINE — BOHRMASCHINE
LATHE — DREHBANK

BLACKSMITH'S TOOLS
SCHMIEDEWERKZEUG

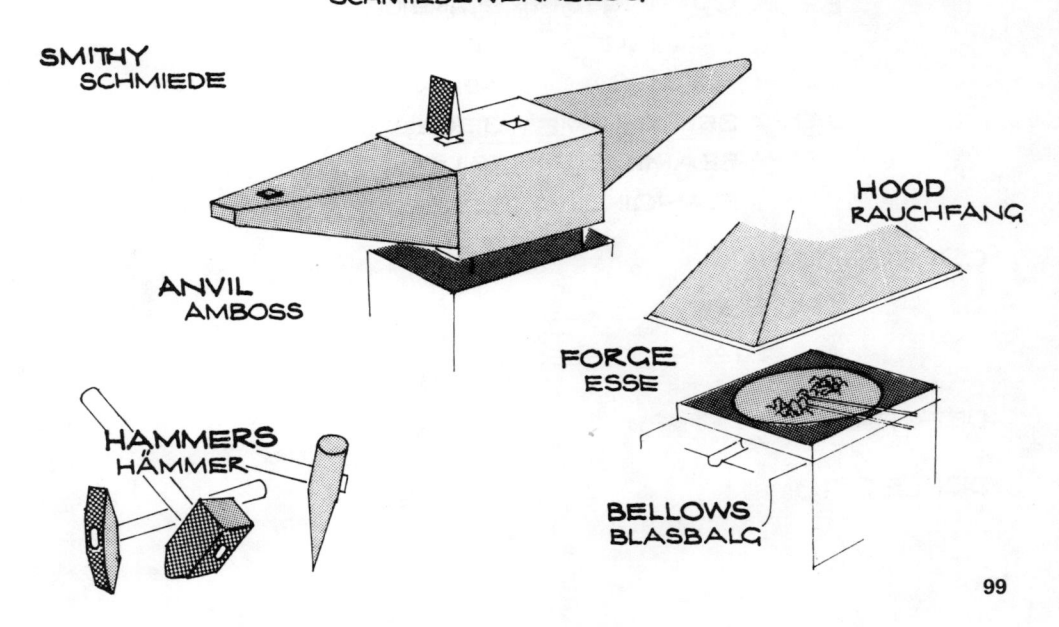

SMITHY — SCHMIEDE
ANVIL — AMBOSS
HOOD — RAUCHFANG
FORGE — ESSE
HAMMERS — HÄMMER
BELLOWS — BLASBALG

TINSMITH'S TOOLS — STEELWORK
SPENGLERWERKZEUG / KLEMPNERWERKZEUG — STAHL- UND METALLARBEITEN

BLOWLAMP — LÖTLAMPE

SOLDERING IRON / GROZING IRON — LÖTKOLBEN

TINSMITH'S SOLDER — LÖTZINN

BRAZING TONGS — LÖTZANGE

TO SOLDER	LÖTEN
TO BRAZE	HARTLÖTEN
TO TIN	VERZINNEN
TO GALVANIZE	VERZINKEN

MACHINES FOR	MASCHINEN ZUM
CUTTING	SCHNEIDEN
DRILLING	BOHREN
FOLDING	ABKANTEN
BENDING	BIEGEN
SEAMING	FALZEN
FLANGING	BÖRDELN

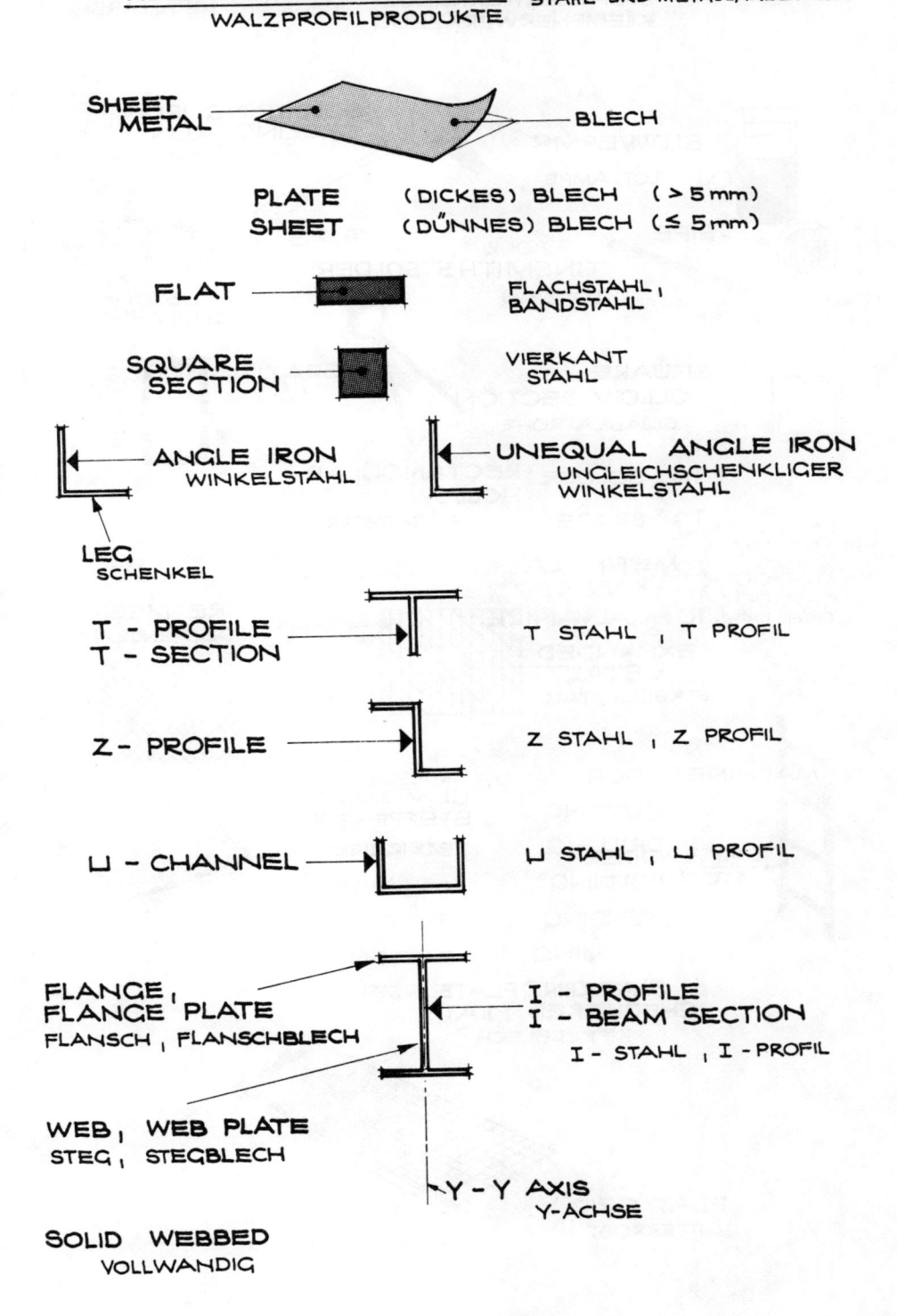

ROLLED STEEL PRODUCTS / WALZPROFILPRODUKTE | STEELWORK / STAHL- UND METALLARBEITEN

- SHEET METAL — BLECH
- PLATE — (DICKES) BLECH (> 5 mm)
- SHEET — (DÜNNES) BLECH (≤ 5 mm)
- FLAT — FLACHSTAHL, BANDSTAHL
- SQUARE SECTION — VIERKANT STAHL
- ANGLE IRON — WINKELSTAHL
- UNEQUAL ANGLE IRON — UNGLEICHSCHENKLIGER WINKELSTAHL
- LEG — SCHENKEL
- T - PROFILE / T - SECTION — T STAHL, T PROFIL
- Z - PROFILE — Z STAHL, Z PROFIL
- U - CHANNEL — U STAHL, U PROFIL
- FLANGE, FLANGE PLATE — FLANSCH, FLANSCHBLECH
- I - PROFILE / I - BEAM SECTION — I - STAHL, I - PROFIL
- WEB, WEB PLATE — STEG, STEGBLECH
- Y - Y AXIS — Y-ACHSE
- SOLID WEBBED — VOLLWANDIG

OTHER STEEL PRODUCTS / STEELWORK
WEITERE STAHLPRODUKTE / STAHL- UND METALLARBEITEN

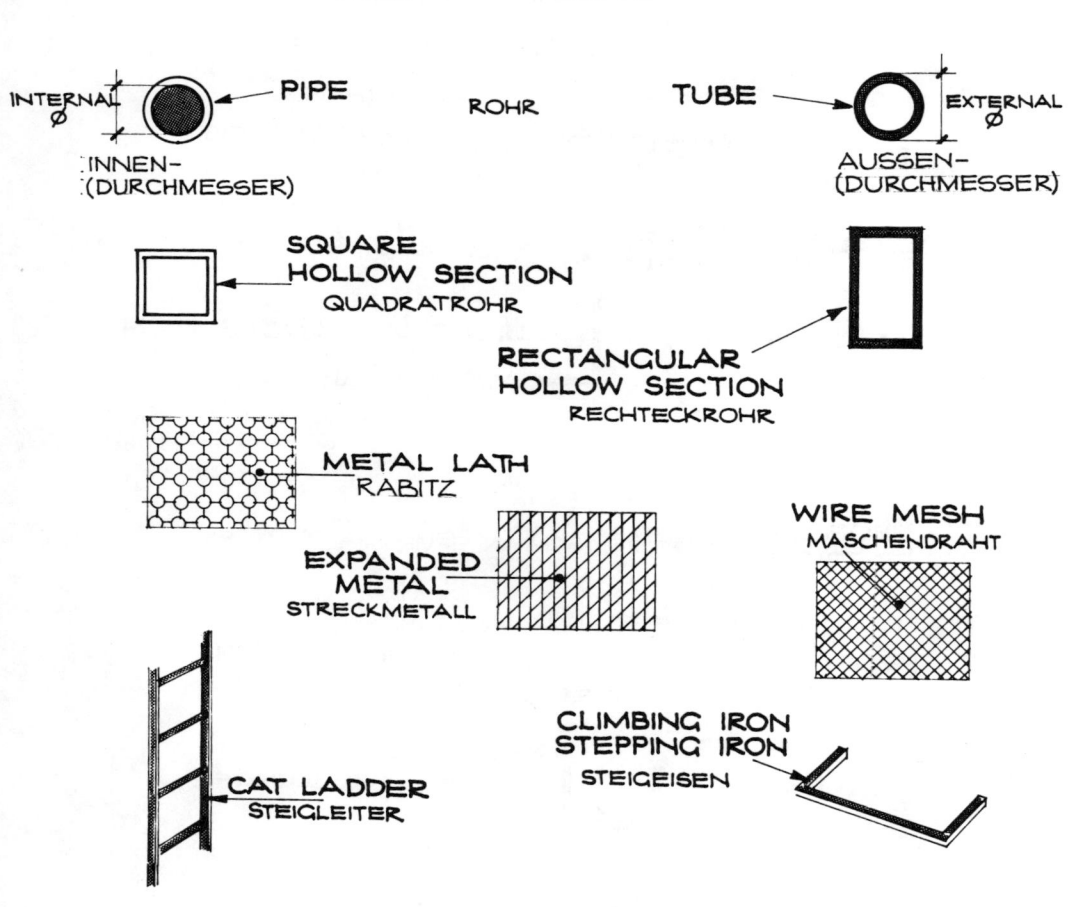

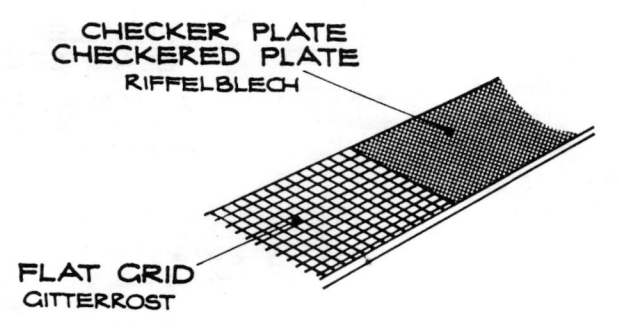

- WIRE — DRAHT
- PIPE — ROHR — INTERNAL Ø — INNEN-(DURCHMESSER)
- TUBE — EXTERNAL Ø — AUSSEN-(DURCHMESSER)
- SQUARE HOLLOW SECTION — QUADRATROHR
- RECTANGULAR HOLLOW SECTION — RECHTECKROHR
- METAL LATH — RABITZ
- EXPANDED METAL — STRECKMETALL
- WIRE MESH — MASCHENDRAHT
- CAT LADDER — STEIGLEITER
- CLIMBING IRON / STEPPING IRON — STEIGEISEN
- CHECKER PLATE / CHECKERED PLATE — RIFFELBLECH
- FLAT GRID — GITTERROST

RAILS
SCHIENEN

STEELWORK
STAHL-UND METALLARBEITEN

RAIL → → ← ← SCHIENE, GLEIS

CRANE RAIL — KRANBAHN
TROLLEY RAIL — LAUFKATZENGLEIS

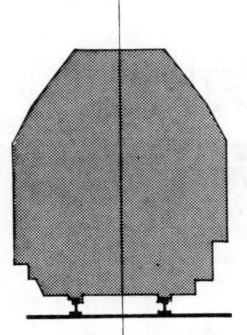

GAUGE
SPURWEITE

GAUGE = ALLG. EICHMASS
GAUGE = MASSEINHEIT FÜR METALLDICKEN

NARROW GAUGE — SCHMALSPUR

STANDARD GAUGE — NORMALSPUR

BROAD GAUGE — BREITSPUR

POINTS — WEICHE

CROSSING — KREUZUNG

RAILWAY LINE
BAHNSTRECKE

SIDING
NEBENGLEIS, GLEISANSCHLUSS

SHUNTING LINE
RANGIERGLEIS

CLEARANCE DIAGRAM
LICHTRAUMPROFIL

METHODS OF JOINTING — STEELWORK
VERBINDUNGSMITTEL — STAHL- UND METALLARBEITEN

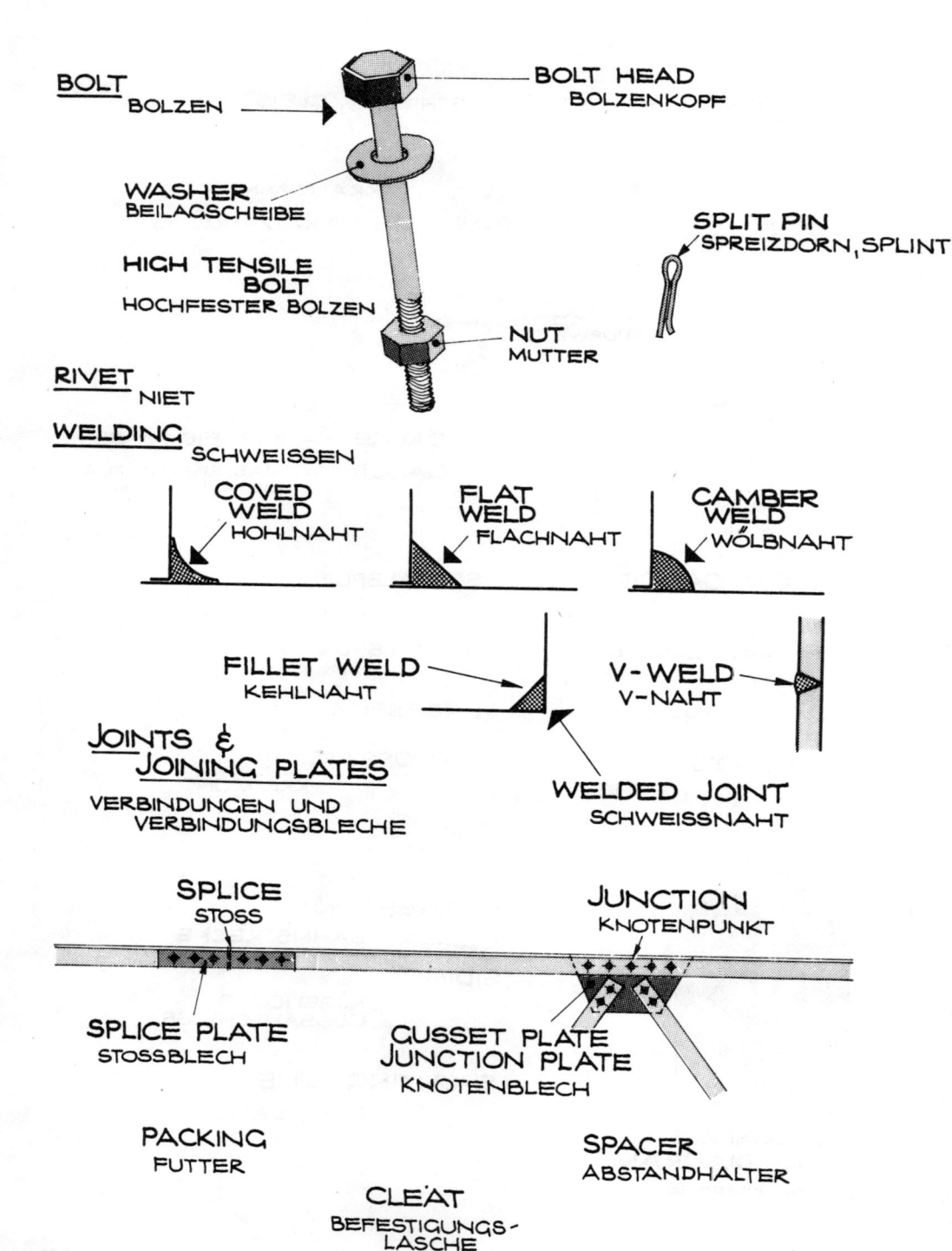

BOLT BOLZEN

BOLT HEAD BOLZENKOPF

WASHER BEILAGSCHEIBE

HIGH TENSILE BOLT HOCHFESTER BOLZEN

NUT MUTTER

SPLIT PIN SPREIZDORN, SPLINT

RIVET NIET

WELDING SCHWEISSEN

COVED WELD HOHLNAHT

FLAT WELD FLACHNAHT

CAMBER WELD WÖLBNAHT

FILLET WELD KEHLNAHT

V-WELD V-NAHT

WELDED JOINT SCHWEISSNAHT

JOINTS & JOINING PLATES VERBINDUNGEN UND VERBINDUNGSBLECHE

SPLICE STOSS

JUNCTION KNOTENPUNKT

SPLICE PLATE STOSSBLECH

GUSSET PLATE / JUNCTION PLATE KNOTENBLECH

PACKING FUTTER

SPACER ABSTANDHALTER

CLEAT BEFESTIGUNGSLASCHE

STEEL STRUCTURES
STAHLTRAGWERKE

STEELWORK
STAHL- UND METALLARBEITEN

GIRDER / TRÄGER   (OBER - UND UNTERGURT SIND PARALLEL)

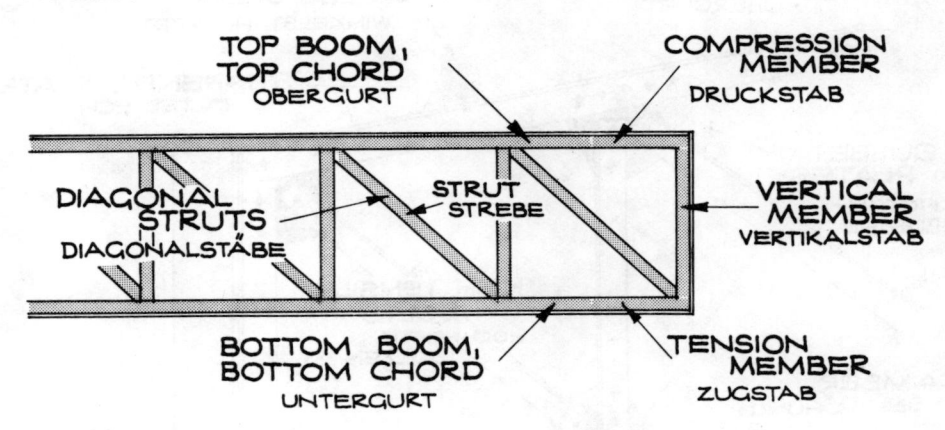

- TOP BOOM, TOP CHORD — OBERGURT
- COMPRESSION MEMBER — DRUCKSTAB
- DIAGONAL STRUTS — DIAGONALSTÄBE
- STRUT — STREBE
- VERTICAL MEMBER — VERTIKALSTAB
- BOTTOM BOOM, BOTTOM CHORD — UNTERGURT
- TENSION MEMBER — ZUGSTAB

LATTICE GIRDER	FACHWERKTRÄGER
HOLLOW GIRDER	HOHLTRÄGER, KASTENTRÄGER
PLATE GIRDER	VOLLWANDTRÄGER
OPEN WEB GIRDER	TRÄGER MIT DURCHBROCHENEM STEG Z.B. WABENTRÄGER, VIERENDEELTRÄGER

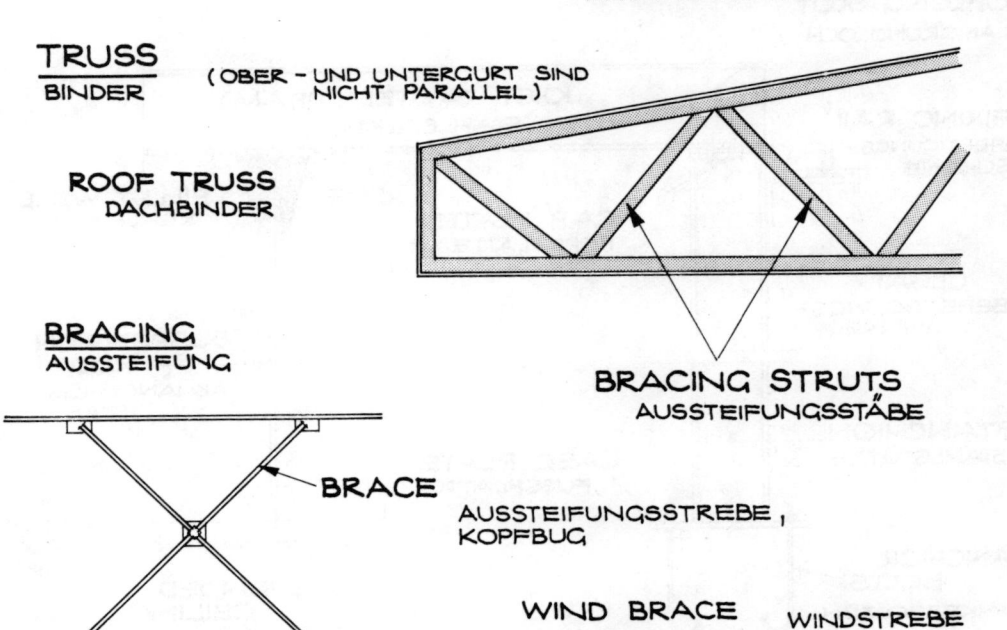

TRUSS / BINDER   (OBER - UND UNTERGURT SIND NICHT PARALLEL)

ROOF TRUSS — DACHBINDER

BRACING / AUSSTEIFUNG

- BRACING STRUTS — AUSSTEIFUNGSSTÄBE
- BRACE — AUSSTEIFUNGSSTREBE, KOPFBUG
- WIND BRACE — WINDSTREBE
- CROSS BRACE — QUERSTREBE

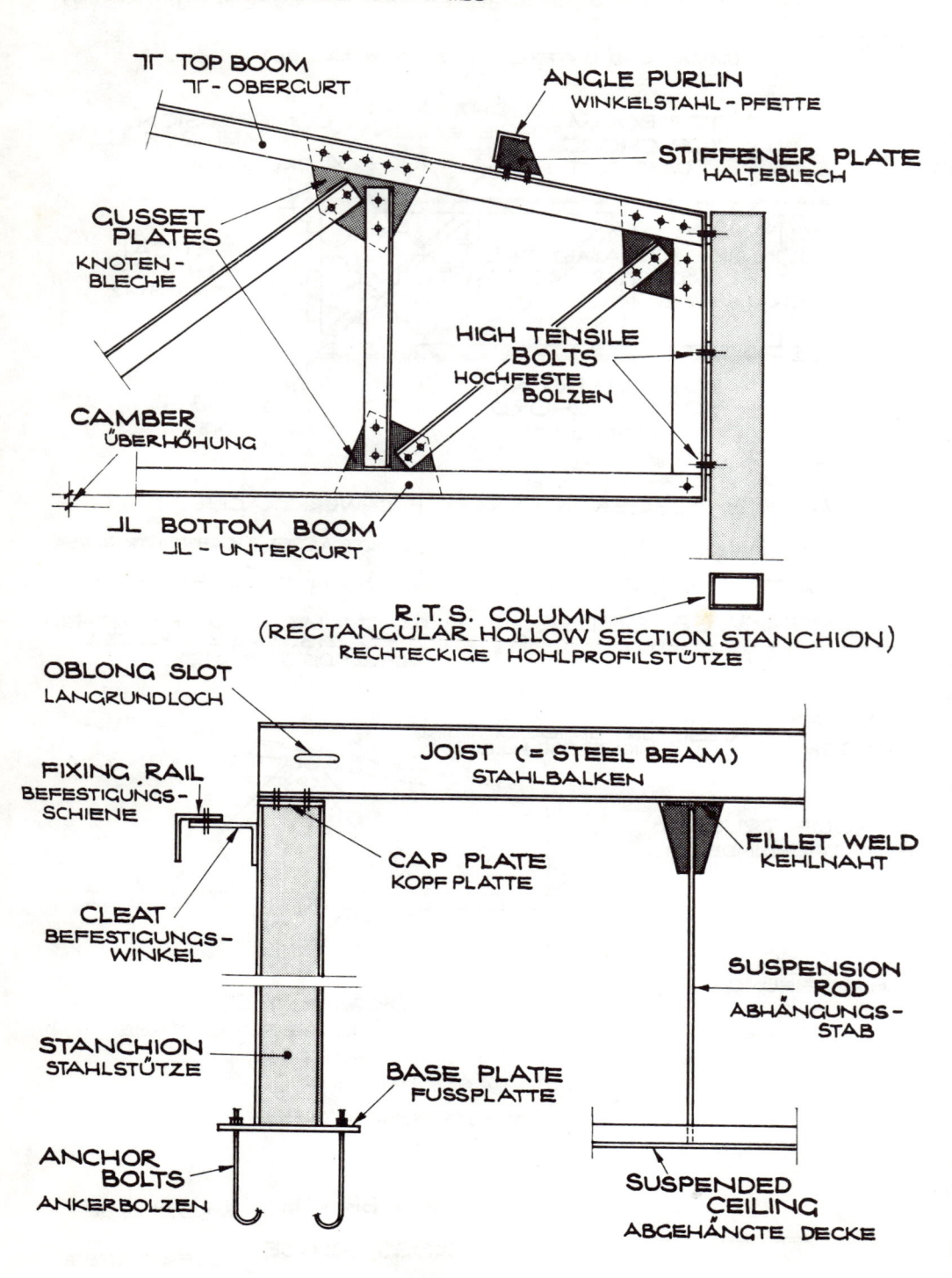

STEELWORK DETAILS / STAHLBAUDETAILS
STEELWORK / STAHL- UND METALLARBEITEN

- ⊤ TOP BOOM / ⊤ - OBERGURT
- ANGLE PURLIN / WINKELSTAHL - PFETTE
- STIFFENER PLATE / HALTEBLECH
- GUSSET PLATES / KNOTENBLECHE
- HIGH TENSILE BOLTS / HOCHFESTE BOLZEN
- CAMBER / ÜBERHÖHUNG
- ⊥ BOTTOM BOOM / ⊥ - UNTERGURT
- R.T.S. COLUMN (RECTANGULAR HOLLOW SECTION STANCHION) / RECHTECKIGE HOHLPROFILSTÜTZE
- OBLONG SLOT / LANGRUNDLOCH
- JOIST (= STEEL BEAM) / STAHLBALKEN
- FIXING RAIL / BEFESTIGUNGSSCHIENE
- CAP PLATE / KOPFPLATTE
- FILLET WELD / KEHLNAHT
- CLEAT / BEFESTIGUNGSWINKEL
- SUSPENSION ROD / ABHÄNGUNGSSTAB
- STANCHION / STAHLSTÜTZE
- BASE PLATE / FUSSPLATTE
- ANCHOR BOLTS / ANKERBOLZEN
- SUSPENDED CEILING / ABGEHÄNGTE DECKE

FORGED STEELWORK
SCHMIEDEARBEITEN

STEELWORK
STAHL- UND METALLARBEITEN

WROUGHT IRON
SCHMIEDEEISEN

ORNAMENTAL WROUGHT IRON WORK
KUNSTSCHMIEDE-ARBEITEN

VOLUTE
SCHNECKE
GEBÄRMUTTER

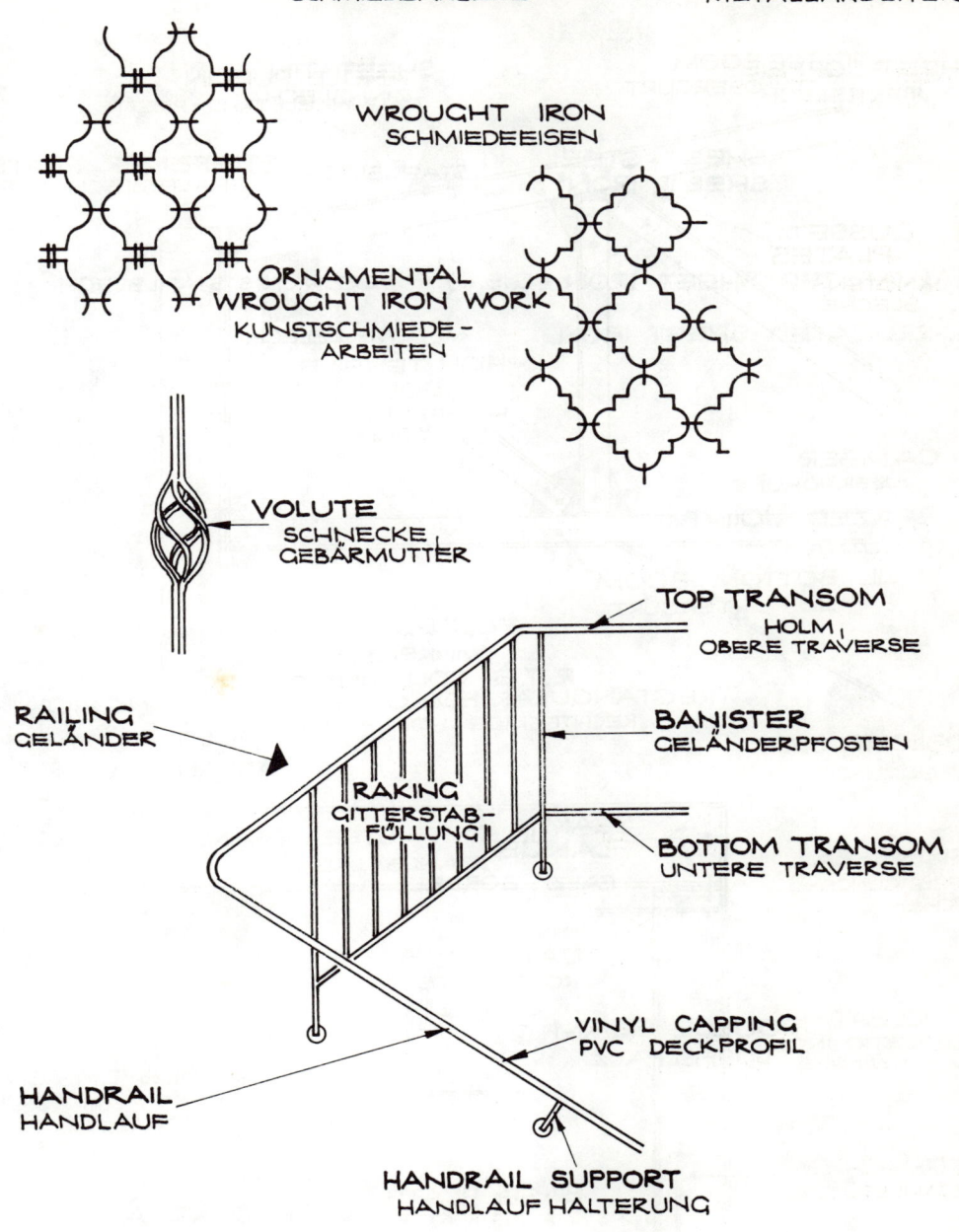

TOP TRANSOM
HOLM, OBERE TRAVERSE

RAILING
GELÄNDER

BANISTER
GELÄNDERPFOSTEN

RAKING
GITTERSTAB-FÜLLUNG

BOTTOM TRANSOM
UNTERE TRAVERSE

VINYL CAPPING
PVC DECKPROFIL

HANDRAIL
HANDLAUF

HANDRAIL SUPPORT
HANDLAUF HALTERUNG

SHEET METALS	STEELWORK
BLECHE	STAHL- UND METALLARBEITEN

SHEET COPPER	SHEET TIN
KUPFERBLECH	ZINNBLECH

SHEET STEEL
SHEET IRON   STAHLBLECH

GALVANIZED SHEET IRON (G.S.I.)   VERZINKTES STAHLBLECH

CORRUGATED SHEET IRON   WELLBLECH

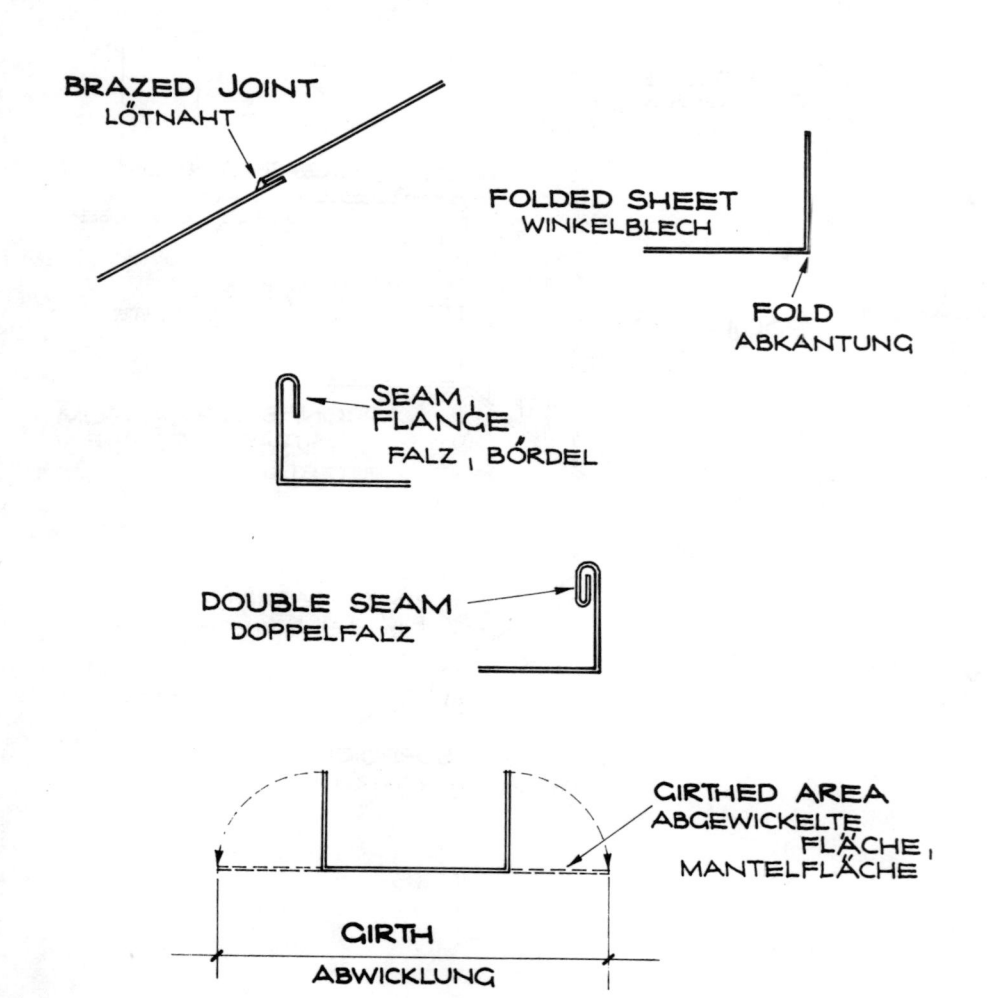

BRAZED JOINT
LÖTNAHT

FOLDED SHEET
WINKELBLECH

FOLD
ABKANTUNG

SEAM
FLANGE
FALZ, BÖRDEL

DOUBLE SEAM
DOPPELFALZ

GIRTHED AREA
ABGEWICKELTE FLÄCHE,
MANTELFLÄCHE

GIRTH
ABWICKLUNG

SHEETMETAL WORK
BLECHARBEITEN

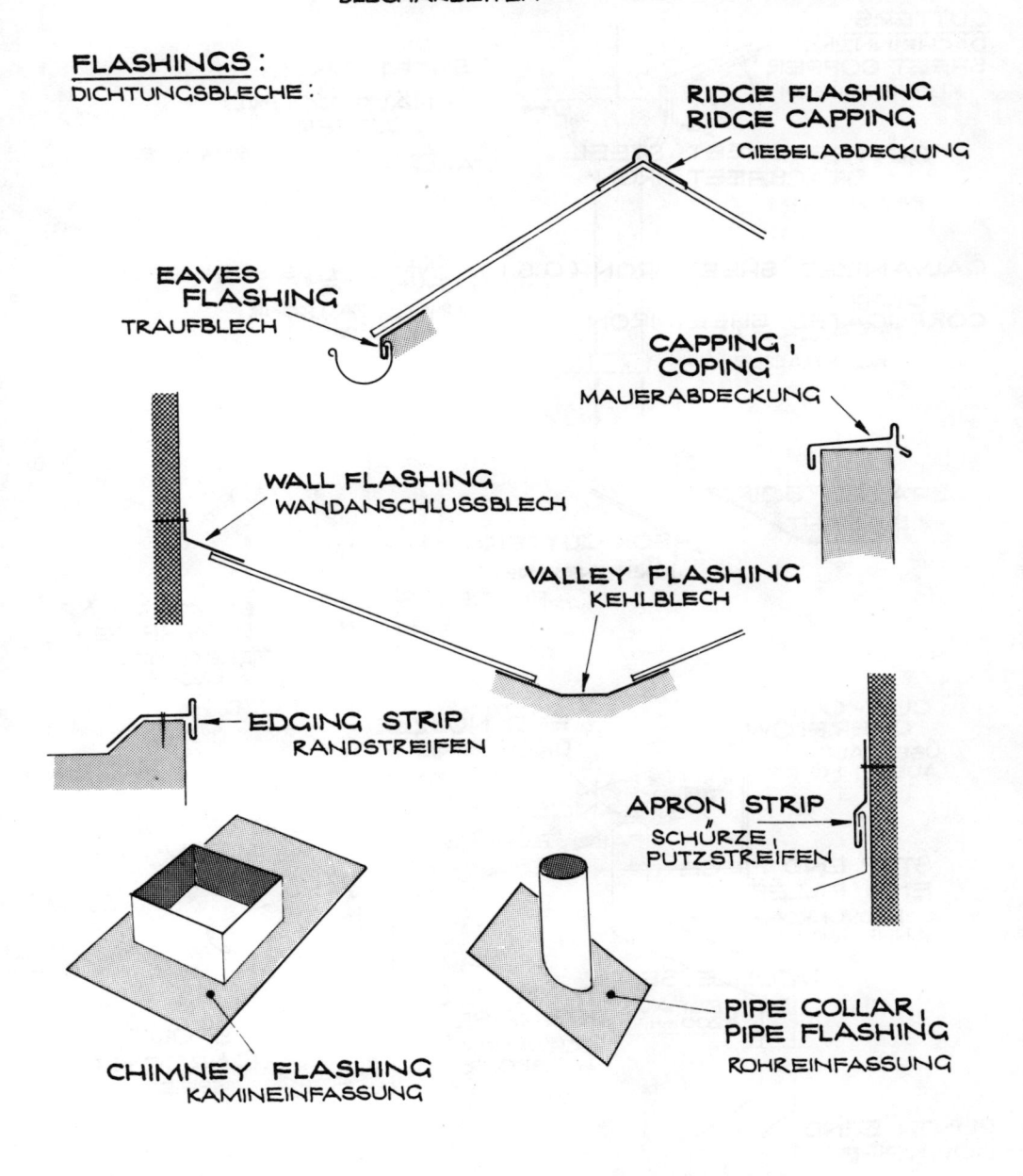

STEELWORK
STAHL- UND METALLARBEITEN

FLASHINGS:
DICHTUNGSBLECHE:

RIDGE FLASHING, RIDGE CAPPING — GIEBELABDECKUNG

EAVES FLASHING — TRAUFBLECH

CAPPING, COPING — MAUERABDECKUNG

WALL FLASHING — WANDANSCHLUSSBLECH

VALLEY FLASHING — KEHLBLECH

EDGING STRIP — RANDSTREIFEN

APRON STRIP — SCHÜRZE, PUTZSTREIFEN

CHIMNEY FLASHING — KAMINEINFASSUNG

PIPE COLLAR, PIPE FLASHING — ROHREINFASSUNG

LINING — AUSKLEIDUNG

EXTERNAL G.S.I. SILL — ÄUSSERES FENSTERBLECH

TINSMITH WORK — STEELWORK
SPENGLER-, KLEMPNERARBEITEN · STAHL- UND METALLARBEITEN

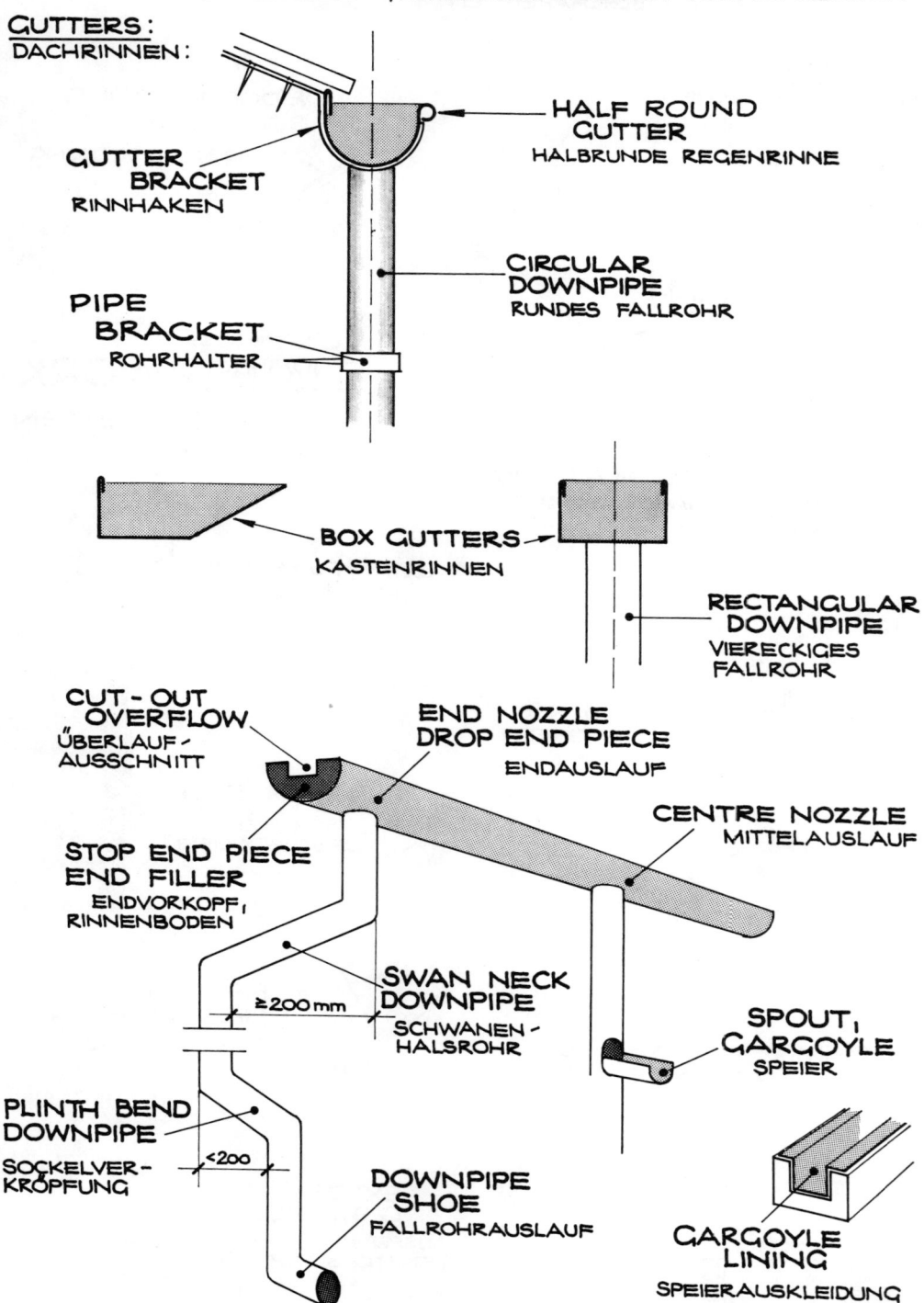

GUTTERS:
DACHRINNEN:

- HALF ROUND GUTTER — HALBRUNDE REGENRINNE
- GUTTER BRACKET — RINNHAKEN
- CIRCULAR DOWNPIPE — RUNDES FALLROHR
- PIPE BRACKET — ROHRHALTER
- BOX GUTTERS — KASTENRINNEN
- RECTANGULAR DOWNPIPE — VIERECKIGES FALLROHR
- CUT-OUT OVERFLOW — ÜBERLAUFAUSSCHNITT
- END NOZZLE, DROP END PIECE — ENDAUSLAUF
- CENTRE NOZZLE — MITTELAUSLAUF
- STOP END PIECE, END FILLER — ENDVORKOPF, RINNENBODEN
- SWAN NECK DOWNPIPE — SCHWANENHALSROHR
- SPOUT, GARGOYLE — SPEIER
- PLINTH BEND DOWNPIPE — SOCKELVERKRÖPFUNG
- DOWNPIPE SHOE — FALLROHRAUSLAUF
- GARGOYLE LINING — SPEIERAUSKLEIDUNG

≥ 200 mm
< 200

TIMBERWORK
HOLZARBEITEN

TIMBERWORK
HOLZARBEITEN

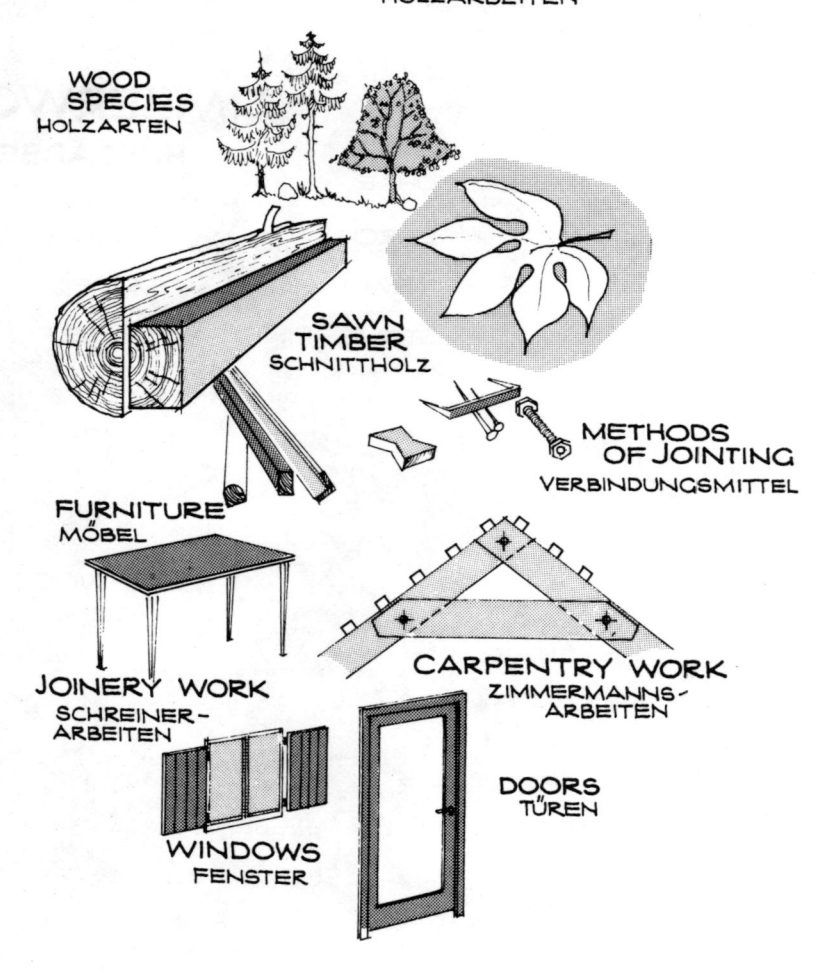

WOOD SPECIES
HOLZARTEN

SAWN TIMBER
SCHNITTHOLZ

METHODS OF JOINTING
VERBINDUNGSMITTEL

FURNITURE
MÖBEL

CARPENTRY WORK
ZIMMERMANNS-ARBEITEN

JOINERY WORK
SCHREINER-ARBEITEN

DOORS
TÜREN

WINDOWS
FENSTER

WOOD SPECIES
HOLZARTEN

TIMBERWORK
HOLZARBEITEN

SOFTWOOD
WEICHHOLZ,
NADELHOLZ

CONIFEROUS WOOD
NADELHOLZ

RESINOUS WOOD
HARZIGES HOLZ

SPRUCE	—	FICHTE
FIR	—	TANNE
PINE	—	KIEFER, FÖHRE
LARCH	—	LÄRCHE
YEW	—	EIBE
STONE-PINE	—	PINIE

HARDWOOD
HARTHOLZ,
LAUBHOLZ

LEAFWOOD
LAUBHOLZ

OAK	—	EICHE
MAPLE	—	AHORN
BEECH	—	BUCHE
BIRCH	—	BIRKE
POPLAR	—	PAPPEL
ELM	—	ULME, RÜSTER
WALNUT	—	NUSSBAUM
ASPEN	—	ESPE
ASH	—	ESCHE
LINDEN	—	LINDE
WILLOW	—	WEIDE
ALDER	—	ERLE

SAWN TIMBER
SCHNITTHOLZ

TIMBERWORK
HOLZARBEITEN

WOOD — WALD, BRENNHOLZ, HOLZ (ALLGEMEIN)

TIMBER — BAUHOLZ

GRADES : KLASSIFIZIERUNG :

UNDERGRADE	UTILITY GRADE	MERCHANTABLE GRADE
AUSSCHUSS	GEBRAUCHSKLASSE	HANDELSKLASSE
XXX	UG	MG
ROUGH EDGED	FULL EDGED	SHARP EDGED
BAUMKANTIG	FEHLKANTIG	SCHARFKANTIG

SQUARED TIMBER , RAIL
KANTHOLZ

BATTEN
LATTE

PLANK
BOHLE

BOARD
BRETT

SHAVINGS, CHIPPINGS — SÄGESPÄNE, HOBELSPÄNE

SAW DUST — SÄGEMEHL

NOTE: DEAL — GESCHNITTENES FICHTENHOLZ

TIMBER PRODUCTS / HOLZPRODUKTE
TIMBERWORK / HOLZARBEITEN

TIMBER BOARDS : HOLZPLATTEN :

PARTICLE BOARD	HOLZFASERPLATTE
CHIPBOARD	SPANPLATTE
BLOCKBOARD	STABPLATTE
SOFTBOARD	WEICHFASERPLATTE
HARDBOARD	HARTFASERPLATTE
FIBRE BOARD	FASERPLATTE

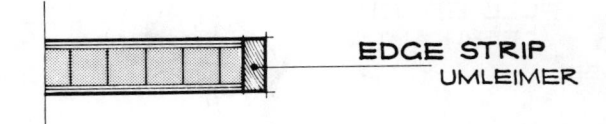

EDGE STRIP — UMLEIMER

VENEER — FURNIER

NOTE:
MAHOGANY — MAHAGONI
AFROMOSIA — AFRORMOSIA

PLYWOOD — SPERRHOLZ

E.G. 5 - PLY
Z.B. 5 - SCHICHTIG

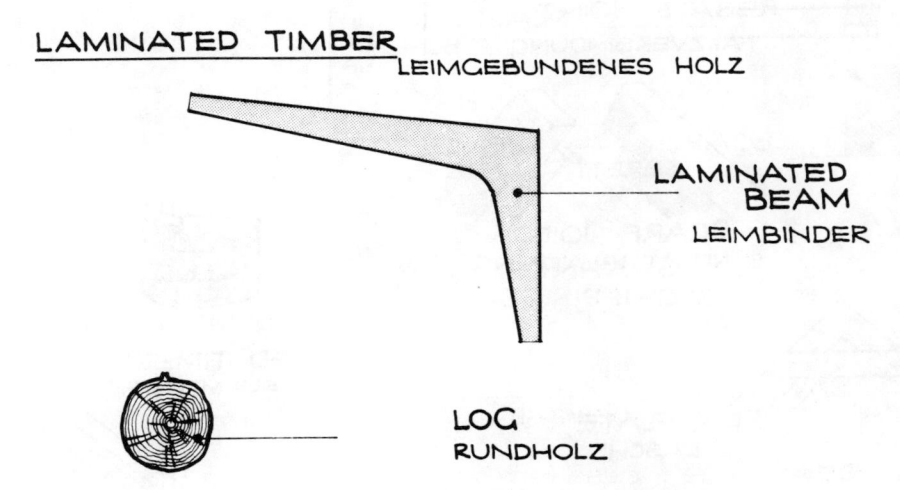

LAMINATED TIMBER — LEIMGEBUNDENES HOLZ

LAMINATED BEAM — LEIMBINDER

LOG — RUNDHOLZ

JOINING MEANS
VERBINDUNGSMITTEL

TIMBERWORK
HOLZARBEITEN

- NAIL — NAGEL
- WOOD SCREW — HOLZSCHRAUBE
- BOLT — BOLZEN
- COACH SCREW — SCHLÜSSELSCHRAUBE
- HARDWOOD PLUG — HARTHOLZDÜBEL
- DOWEL, PEG, GANGNAIL = GANGNAIL, DÜBEL
- TONGUE AND GROOVE — NUT UND FEDER
- MORTICE AND TENON — ZAPFEN UND ZAPFENLOCH
- REBATE JOINT — FALZVERBINDUNG
- SCARF JOINT — BLATT, VERBLATTUNG
- GLUE — LEIM
- FISH PLATE — LASCHE
- GLUED TIMBER — GELEIMTES HOLZ
- SPLICE PIECE — LASCHE

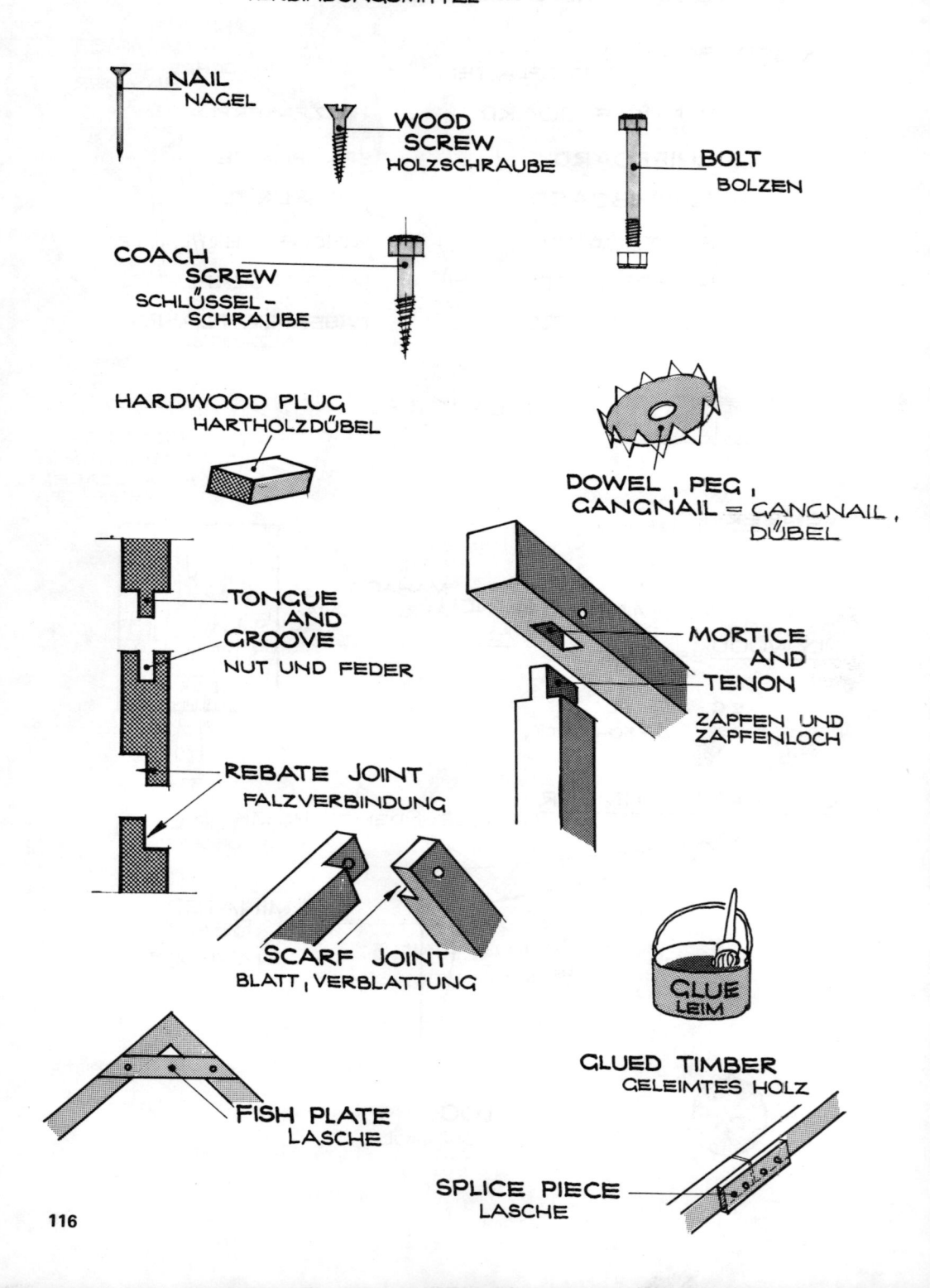

116

JOINER'S TOOLS
SCHREINERWERKZEUG

TIMBERWORK
HOLZARBEITEN

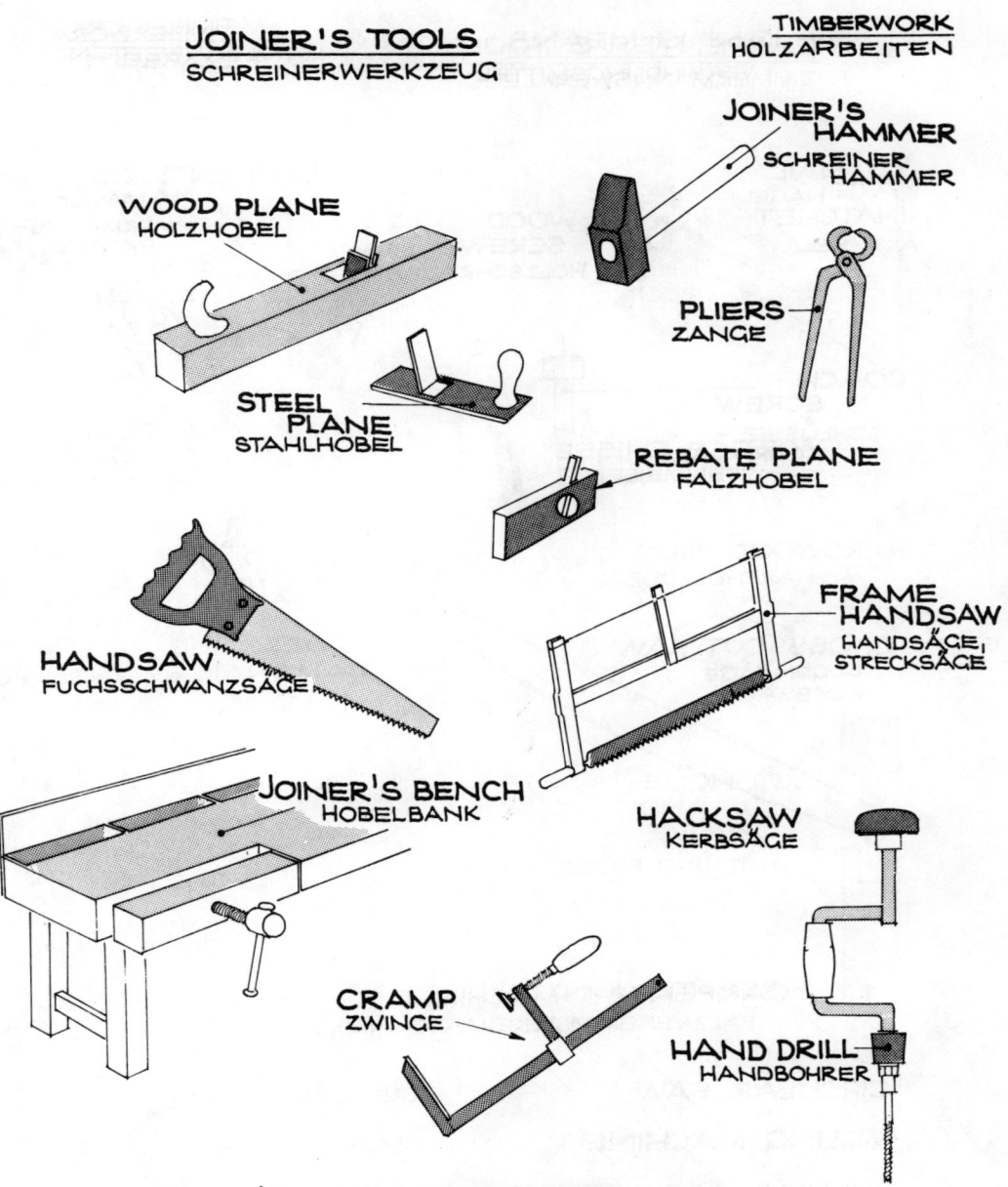

WOOD PLANE
HOLZHOBEL

JOINER'S HAMMER
SCHREINER HAMMER

PLIERS
ZANGE

STEEL PLANE
STAHLHOBEL

REBATE PLANE
FALZHOBEL

HANDSAW
FUCHSSCHWANZSÄGE

FRAME HANDSAW
HANDSÄGE, STRECKSÄGE

JOINER'S BENCH
HOBELBANK

HACKSAW
KERBSÄGE

CRAMP
ZWINGE

HAND DRILL
HANDBOHRER

JOINERY MACHINES:
SCHREINEREIMASCHINEN:

BAND SAW	BANDSÄGE
PLANING MACHINE	HOBELMASCHINE, ABRICHTE
LATHE	DREHBANK
VENEERING PRESS	FURNIERPRESSE

117

CARPENTER'S TOOLS
ZIMMERMANNSWERKZEUG

TIMBERWORK
HOLZARBEITEN

AXE, HATCHET
AXT, BEIL

CARPENTER'S HAMMER
ZIMMERMANNS-HAMMER

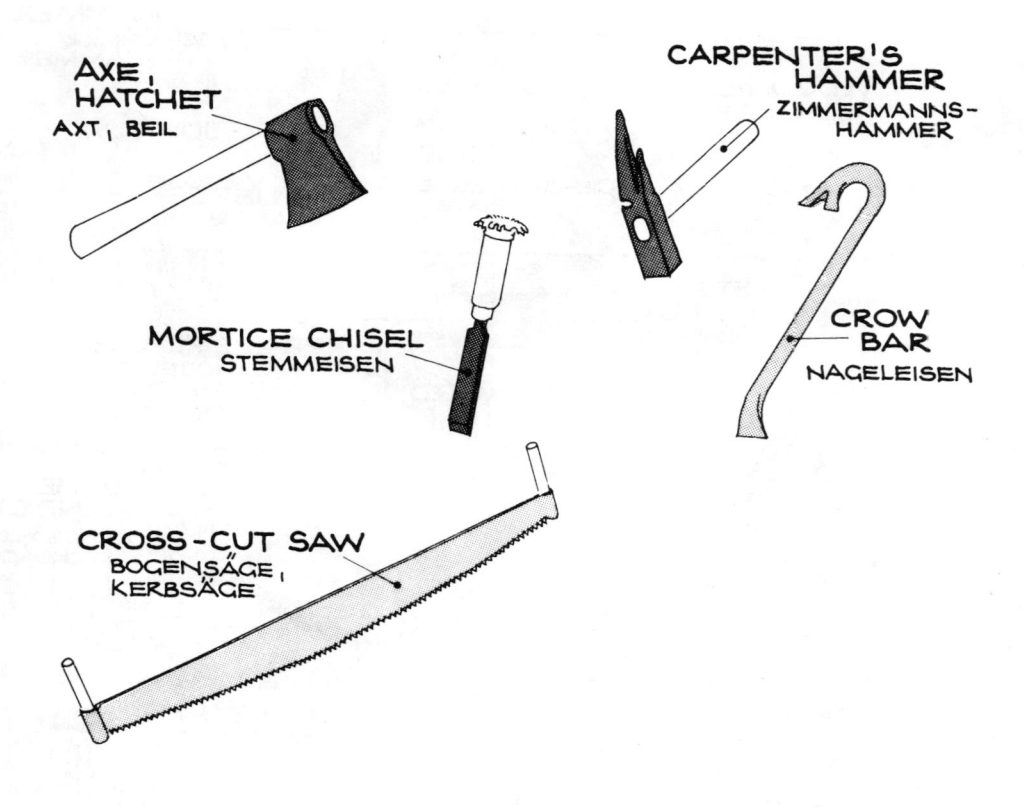

MORTICE CHISEL
STEMMEISEN

CROW BAR
NAGELEISEN

CROSS-CUT SAW
BOGENSÄGE, KERBSÄGE

CARPENTRY MACHINES:
ZIMMEREIMASCHINEN:

CIRCULAR SAW	KREISSÄGE
MILLING MACHINE	FRÄSE
DRILLING MACHINE	BOHRMASCHINE
GRINDER	SCHLEIFMASCHINE
GANG SAW	VOLLGATTER

JOINERY DETAILS / TIMBERWORK
SCHREINERDETAILS / HOLZARBEITEN

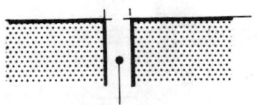

OPEN JOINT — OFFENE FUGE 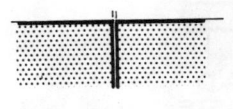
CLOSED JOINT — GESCHLOSSENE FUGE 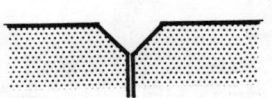
V-JOINT — V-FUGE

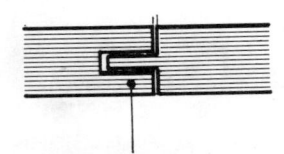

TONGUED AND GROOVED — GENUTET UND GEFEDERT 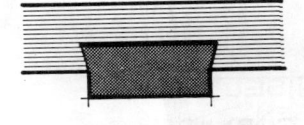
DOVETAILED — KONISCH EINGENUTET 
DOVETAIL DOWEL — SCHWALBENSCHWANZ-DÜBEL

DOVETAIL JOINT — VERZINKUNG
TENON AND MORTICE — ZAPFEN UND ZAPFENLOCH

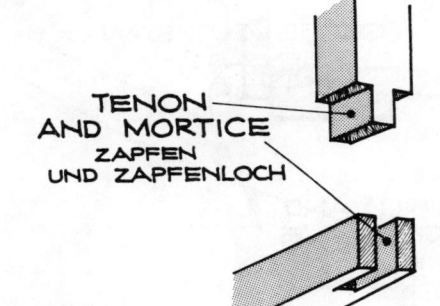

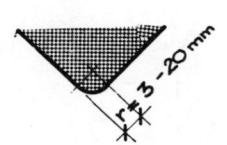

ANGLE ROUNDED — ABGERUNDETE KANTEN (r = 3–20 mm)
ARRIS ROUNDED — GEBROCHENE KANTEN

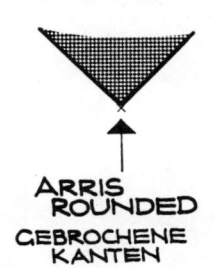

TO SAW	SÄGEN
TO PLANE	HOBELN
TO SANDPAPER	SCHLEIFEN
TO PIN	ANHEFTEN
TO WEDGE	VERKEILEN
TO KEY	EINKEILEN
TO CLAMP	FESTKLAMMERN
TO SCRIBE	ANREISSEN, MARKIEREN
TO FIT	ANPASSEN
TO GLUE	LEIMEN
TO MITRE	AUF GEHRUNG SCHNEIDEN

BEADS AND FILLETS
LEISTEN

TIMBERWORK
HOLZARBEITEN

BATTEN
LATTE

BEAD
LEISTE

FILLET
EINLEGLEISTE

SKIRTING
SOCKELLEISTE

CORNICE
ECKLEISTE

JAMB
FUTTERLEISTE

LINING AND ARCHITRAVE
FUTTER UND VERKLEIDUNG

EDGE PROTECTION BEAD
ECKSCHUTZLEISTE

RAIL
KANTHOLZ, BALKEN

STRINGER
TREPPENWANGE

SPIRAL STRINGER
KRÜMMER

FURNITURE
MÖBEL

TIMBERWORK
HOLZARBEITEN

CABINET-MAKER
MÖBELSCHREINER

FITMENTS :
EINRICHTUNGSGEGENSTÄNDE :

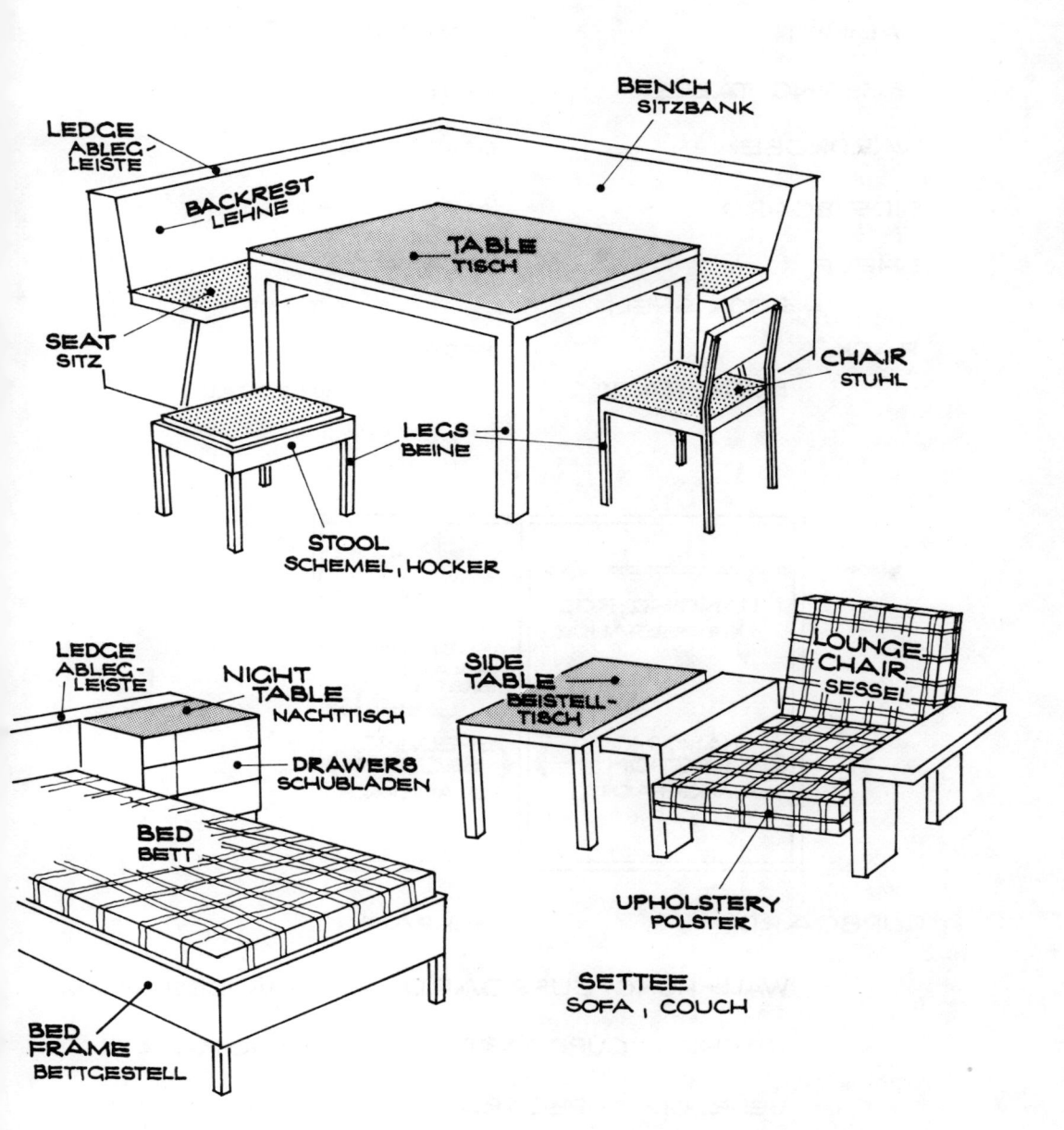

- BENCH — SITZBANK
- LEDGE — ABLEGLEISTE
- BACKREST — LEHNE
- TABLE — TISCH
- SEAT — SITZ
- CHAIR — STUHL
- LEGS — BEINE
- STOOL — SCHEMEL, HOCKER
- LEDGE — ABLEGLEISTE
- NIGHT TABLE — NACHTTISCH
- DRAWERS — SCHUBLADEN
- BED — BETT
- BED FRAME — BETTGESTELL
- SIDE TABLE — BEISTELLTISCH
- LOUNGE CHAIR — SESSEL
- UPHOLSTERY — POLSTER
- SETTEE — SOFA, COUCH

121

FURNITURE
MÖBEL

TIMBERWORK
HOLZARBEITEN

CABINETS AND CUPBOARDS:
   SCHRÄNKE:

CABINET	GLASSCHRANK, VITRINE
DRESSING TABLE	FRISIERTISCH, KOMMODE
WARDROBE	GARDEROBE
SIDE BOARD	ANRICHTE, BÜFFET
SHELF	ABLAGE
BOOK SHELF	BÜCHERBRETT
RACK	REGAL
BOOK RACK	BÜCHERREGAL

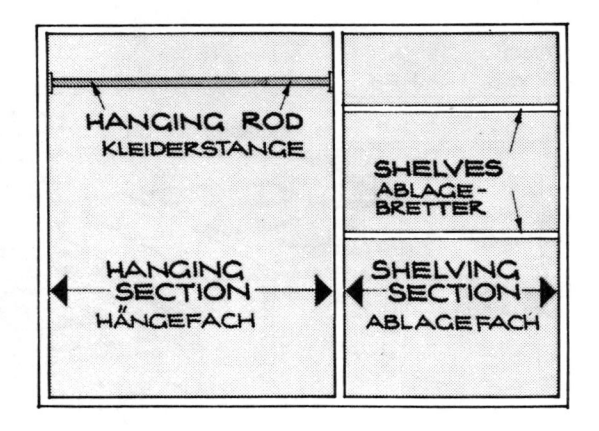

[Diagram of a cupboard with labeled sections:
- HANGING ROD / KLEIDERSTANGE
- SHELVES / ABLAGEBRETTER
- HANGING SECTION / HÄNGEFACH
- SHELVING SECTION / ABLAGEFACH]

CUPBOARD	SCHRANK
WALL-HUNG CUPBOARD	WANDHÄNGESCHRANK
KITCHEN CUPBOARD	KÜCHENSCHRANK
BEDROOM CUPBOARD	SCHLAFZIMMERSCHRANK
BUILT-IN CUPBOARD	EINBAUSCHRANK

FURNITURE DETAILS

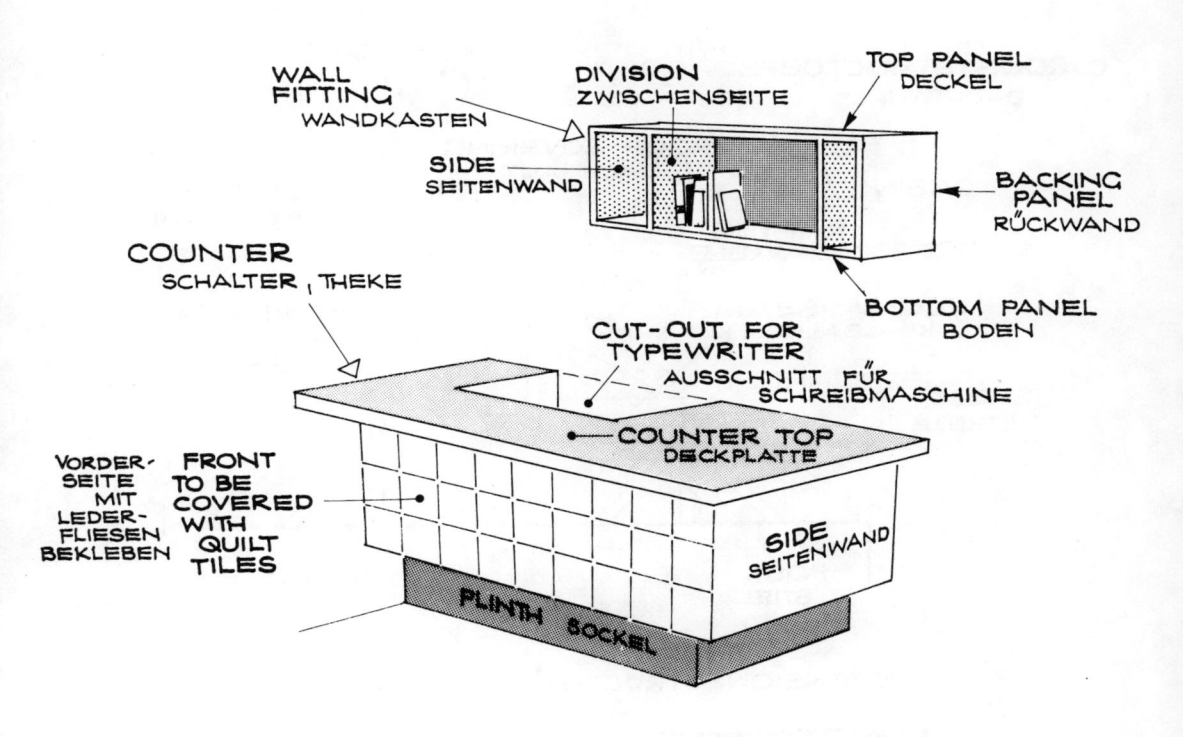

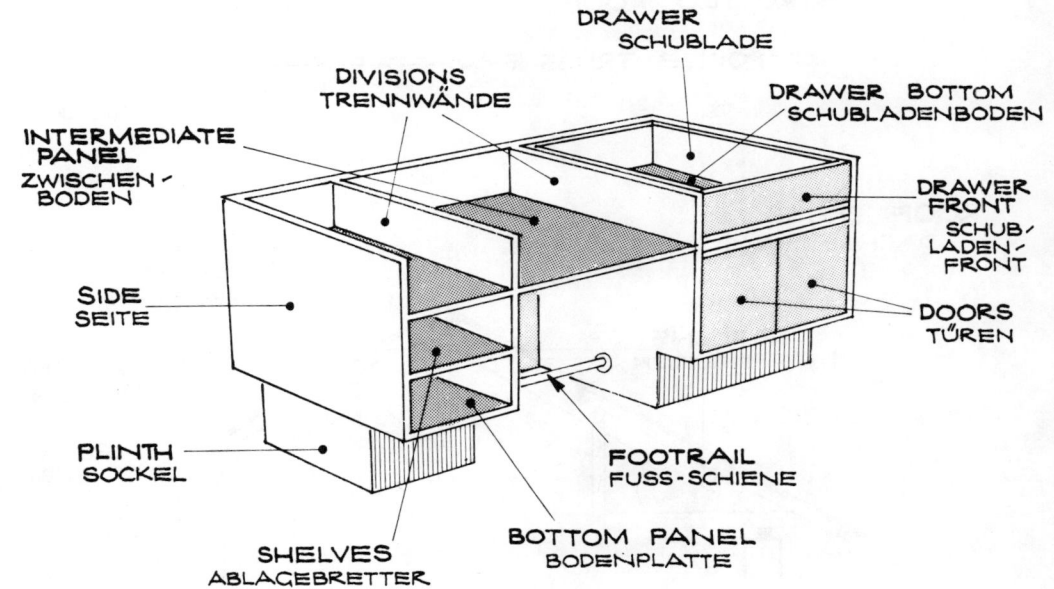

TIMBERWORK
HOLZARBEITEN

- WALL FITTING — WANDKASTEN
- DIVISION — ZWISCHENSEITE
- TOP PANEL — DECKEL
- SIDE — SEITENWAND
- BACKING PANEL — RÜCKWAND
- BOTTOM PANEL — BODEN

- COUNTER — SCHALTER, THEKE
- CUT-OUT FOR TYPEWRITER — AUSSCHNITT FÜR SCHREIBMASCHINE
- COUNTER TOP — DECKPLATTE
- FRONT TO BE COVERED WITH QUILT TILES — VORDERSEITE MIT LEDERFLIESEN BEKLEBEN
- SIDE — SEITENWAND
- PLINTH — SOCKEL

- DRAWER — SCHUBLADE
- DRAWER BOTTOM — SCHUBLADENBODEN
- DIVISIONS — TRENNWÄNDE
- INTERMEDIATE PANEL — ZWISCHENBODEN
- DRAWER FRONT — SCHUBLADENFRONT
- SIDE — SEITE
- DOORS — TÜREN
- PLINTH — SOCKEL
- FOOTRAIL — FUSS-SCHIENE
- SHELVES — ABLAGEBRETTER
- BOTTOM PANEL — BODENPLATTE

123

CARPENTER WORK
ZIMMERMANNSARBEITEN

TIMBERWORK
HOLZARBEITEN

ROOF STRUCTURE:
DACHSTUHL:

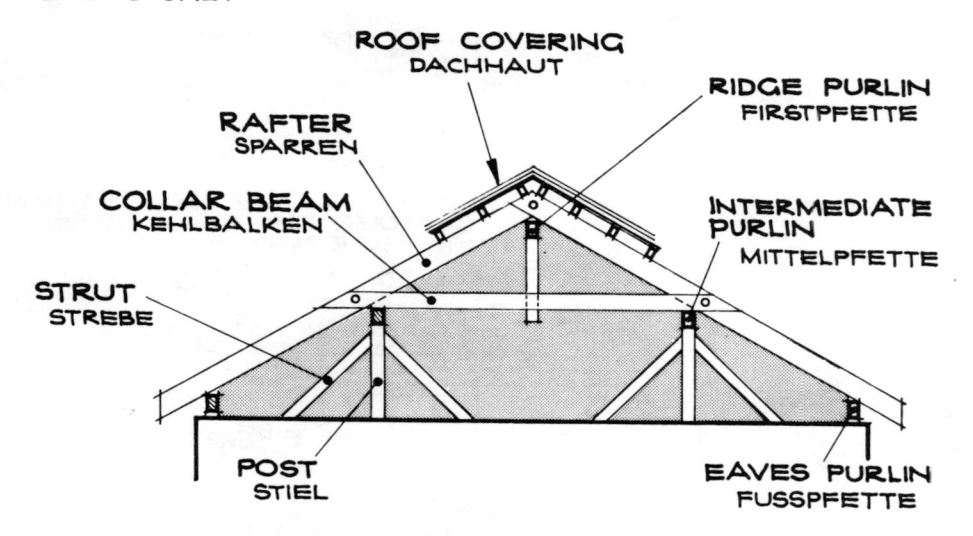

- ROOF COVERING — DACHHAUT
- RIDGE PURLIN — FIRSTPFETTE
- RAFTER — SPARREN
- COLLAR BEAM — KEHLBALKEN
- INTERMEDIATE PURLIN — MITTELPFETTE
- STRUT — STREBE
- POST — STIEL
- EAVES PURLIN — FUSSPFETTE

SUSPENSION STRUCTURE — HÄNGEWERK
STRUTTED ROOF — SPRENGWERK
COMPOSITE TRUSS FRAMED ROOF — HÄNGE- UND SPRENGWERK

ROOF TRUSS:
DACHBINDER:

- TOP CHORD / TOP BOOM — OBERGURT
- PURLINS — PFETTEN
- KING POST — HAUPTSTIEL
- POST — STIEL
- STRUTS — STREBEN
- BOTTOM CHORD / BOTTOM BOOM — UNTERGURT

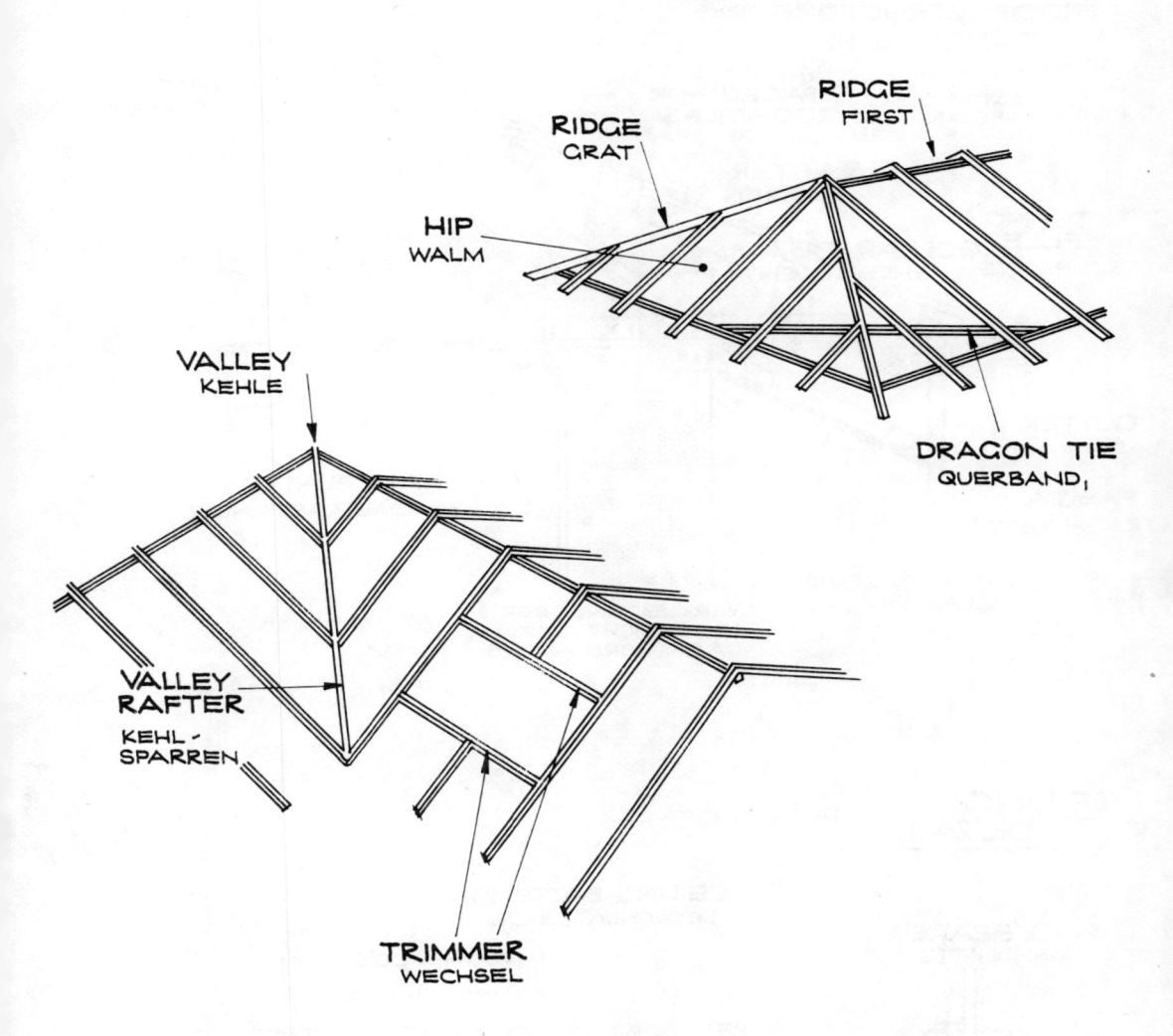

CARPENTER WORK
ZIMMERMANNSARBEITEN

TIMBERWORK
HOLZARBEITEN

RIDGE GRAT

RIDGE FIRST

HIP WALM

DRAGON TIE QUERBAND

VALLEY KEHLE

VALLEY RAFTER KEHLSPARREN

TRIMMER WECHSEL

CARPENTER'S DETAILS
ZIMMERMANNSDETAILS

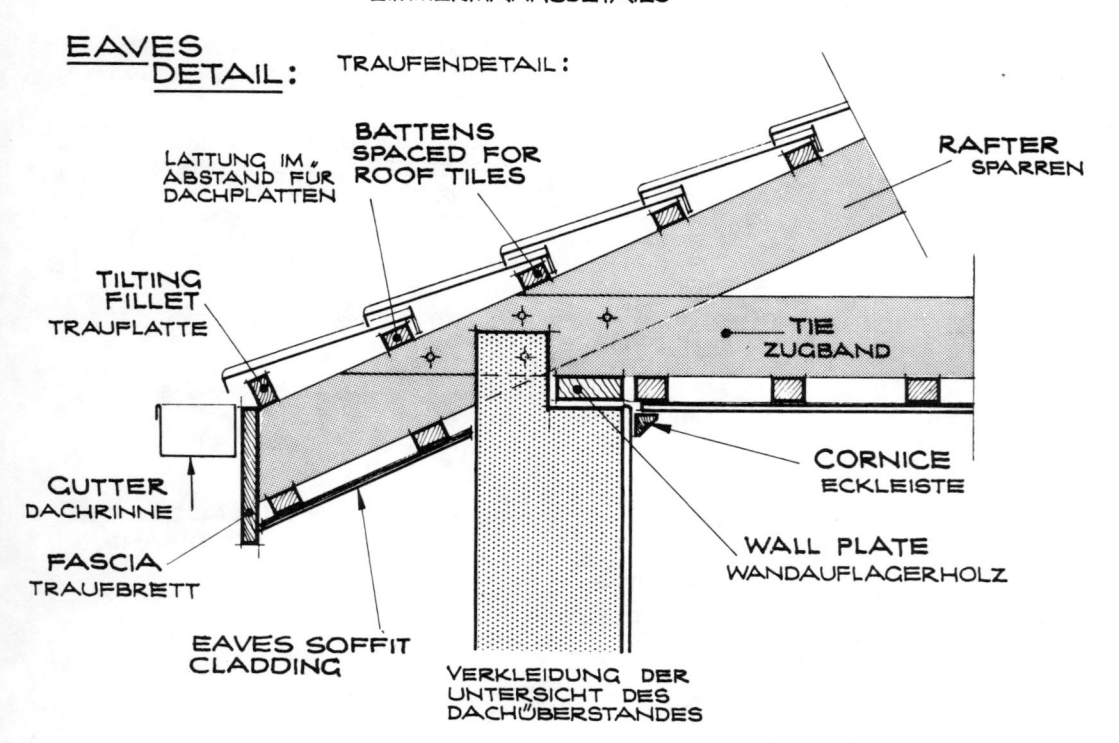

TIMBERWORK
HOLZARBEITEN

EAVES DETAIL:
TRAUFENDETAIL:

- LATTUNG IM ABSTAND FÜR DACHPLATTEN
- **BATTENS SPACED FOR ROOF TILES**
- **RAFTER** / SPARREN
- **TILTING FILLET** / TRAUFLATTE
- **TIE** / ZUGBAND
- **GUTTER** / DACHRINNE
- **CORNICE** / ECKLEISTE
- **FASCIA** / TRAUFBRETT
- **WALL PLATE** / WANDAUFLAGERHOLZ
- **EAVES SOFFIT CLADDING** / VERKLEIDUNG DER UNTERSICHT DES DACHÜBERSTANDES

CEILING DETAIL:
DECKENDETAIL:

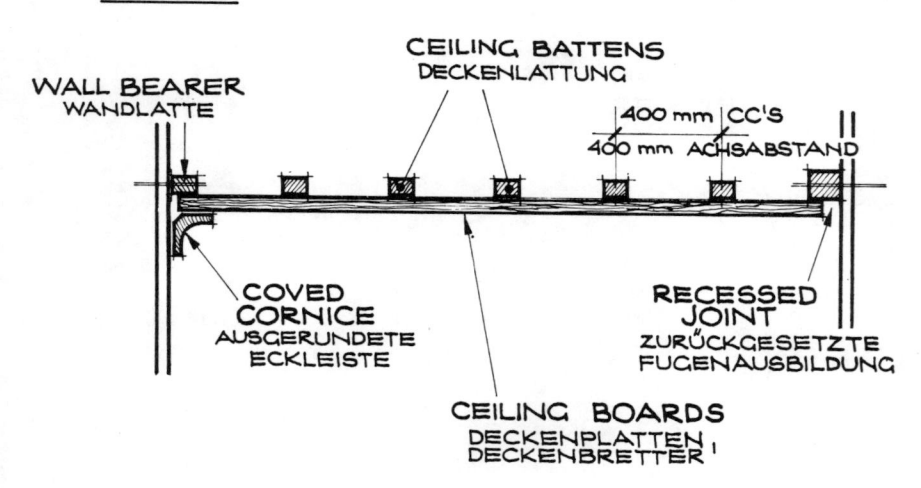

- **WALL BEARER** / WANDLATTE
- **CEILING BATTENS** / DECKENLATTUNG
- 400 mm CC'S / 400 mm ACHSABSTAND
- **COVED CORNICE** / AUSGERUNDETE ECKLEISTE
- **RECESSED JOINT** / ZURÜCKGESETZTE FUGENAUSBILDUNG
- **CEILING BOARDS** / DECKENPLATTEN / DECKENBRETTER

TIMBER WINDOWS
HOLZFENSTER

TIMBERWORK
HOLZARBEITEN

BEACHTE: IN DEN ENGLISCHSPRECHENDEN LÄNDERN DER WARMEN KLIMAZONEN (AUSTRALIEN, SÜDAFRIKA USW.) SIND HOLZFENSTER NAHEZU UNGEBRÄUCHLICH.

STATTDESSEN WERDEN FENSTER AUS STAHLPROFIL-RAHMEN, DIE IN GENORMTEN GRÖSSEN UND TYPEN HERGESTELLT WERDEN VERWENDET.

NOTE: IN ENGLISH SPEAKING COUNTRIES WITH WARM CLIMATES (AUSTRALIA, SOUTHAFRICA ETC.) TIMBER WINDOWS ARE VERY UNCOMMON.

INSTEAD OF THESE, THEY INSTALL WINDOWS WITH STEEL-FRAMES IN STANDARD SIZES AND TYPES.

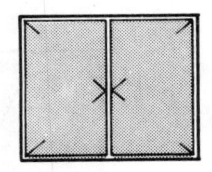

SINGLE WINDOW
EINZELFENSTER

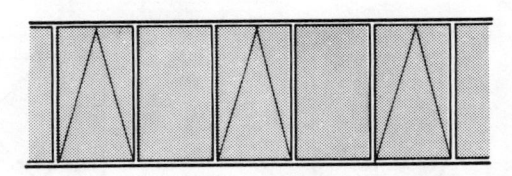

CONTINUOUS WINDOW
FENSTERBAND

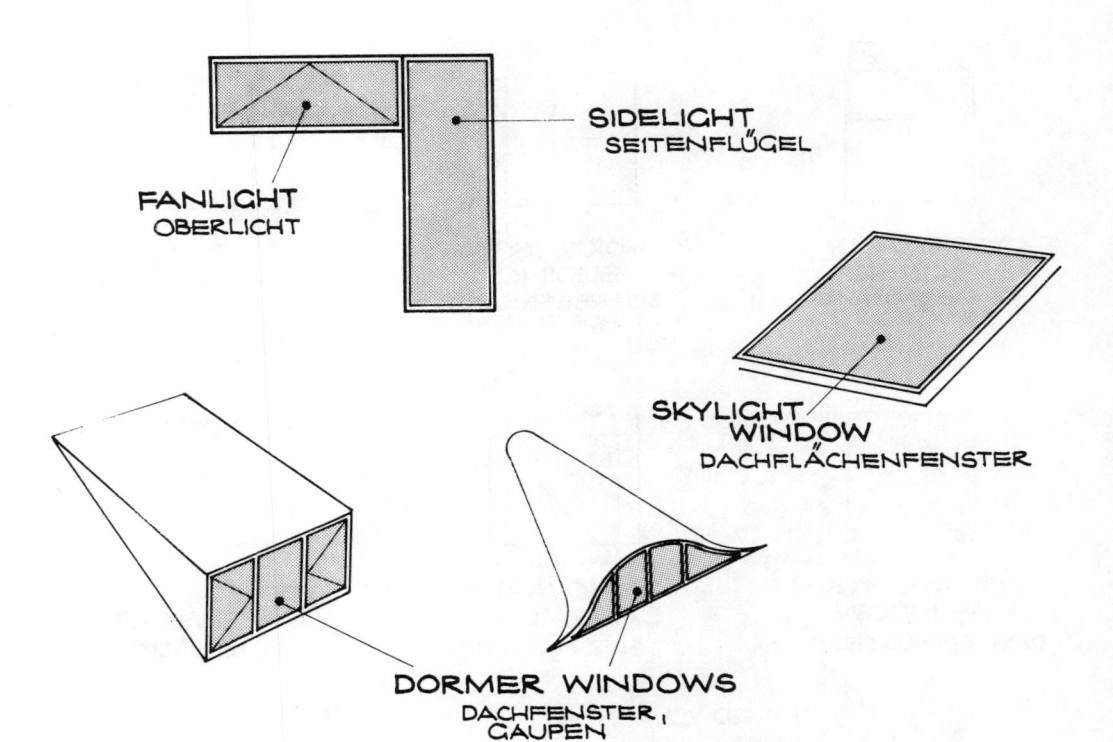

FANLIGHT
OBERLICHT

SIDELIGHT
SEITENFLÜGEL

SKYLIGHT WINDOW
DACHFLÄCHENFENSTER

DORMER WINDOWS
DACHFENSTER, GAUPEN

WINDOW OPENING TYPES
FENSTERÖFFNUNGSARTEN

TIMBERWORK
HOLZARBEITEN

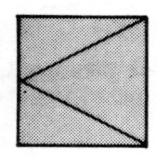

SIDE HUNG RIGHT HAND
DREHFLÜGEL DIN RECHTS

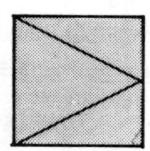

SIDE HUNG LEFT HAND
DREHFLÜGEL DIN LINKS

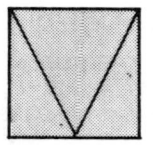

TOP HUNG
KLAPPFLÜGEL

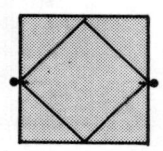

HORIZONTALLY PIVOT HUNG
SCHWINGFLÜGEL

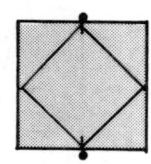

VERTICALLY PIVOTTED
WENDEFLÜGEL

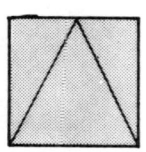

BOTTOM HUNG HOPPER HUNG
KIPPFLÜGEL

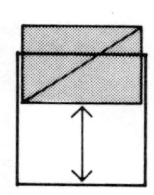

VERTICALLY SLIDING
SCHIEBEFENSTER VERTIKAL

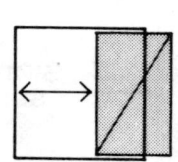

HORIZONTALLY SLIDING
SCHIEBEFENSTER HORIZONTAL

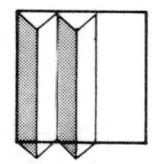

FOLDING
FALTFENSTER

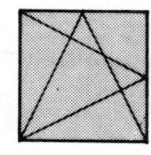

TILT AND TURN WINDOW
DREHKIPPFLÜGEL

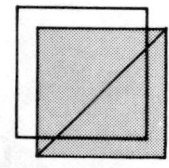

REMOVABLE CASEMENT
ABSTELLFLÜGEL

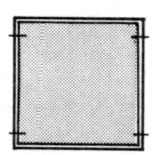

FIXED CASEMENT
PUTZFLÜGEL

FIXED GLAZING
FESTVERGLASUNG

WINDOW SUBDIVISIONS
FENSTERTEILUNG

TIMBERWORK
HOLZARBEITEN

2 SIDE HUNG SECTIONS
2 DREHFLÜGEL

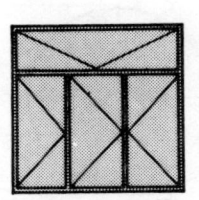

3 SIDE HUNG SECTIONS & 1 TOP HUNG FANLIGHT
3 DREHFLÜGEL UND 1 KLAPPFLÜGEL-OBERLICHT

2 SIDE HUNG & 1 LARGE FIXED SECTION
2 DREHFLÜGEL UND 1 GROSSER FESTVERGLASTER TEIL

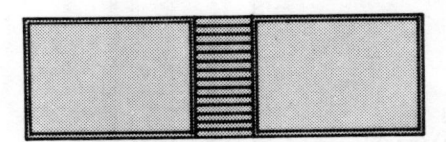

DISPLAY WINDOWS WITH 1 LOUVRED PANEL
SCHAUFENSTER MIT 1 LÜFTUNGSTEIL

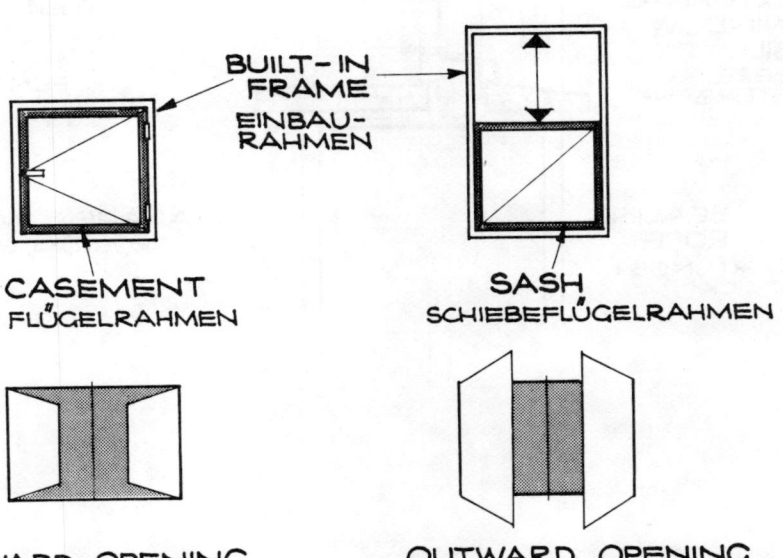

BUILT-IN FRAME
EINBAURAHMEN

CASEMENT
FLÜGELRAHMEN

SASH
SCHIEBEFLÜGELRAHMEN

INWARD OPENING
NACH INNEN ÖFFNEND

OUTWARD OPENING
NACH AUSSEN ÖFFNEND

WINDOW DETAIL
FENSTERDETAIL

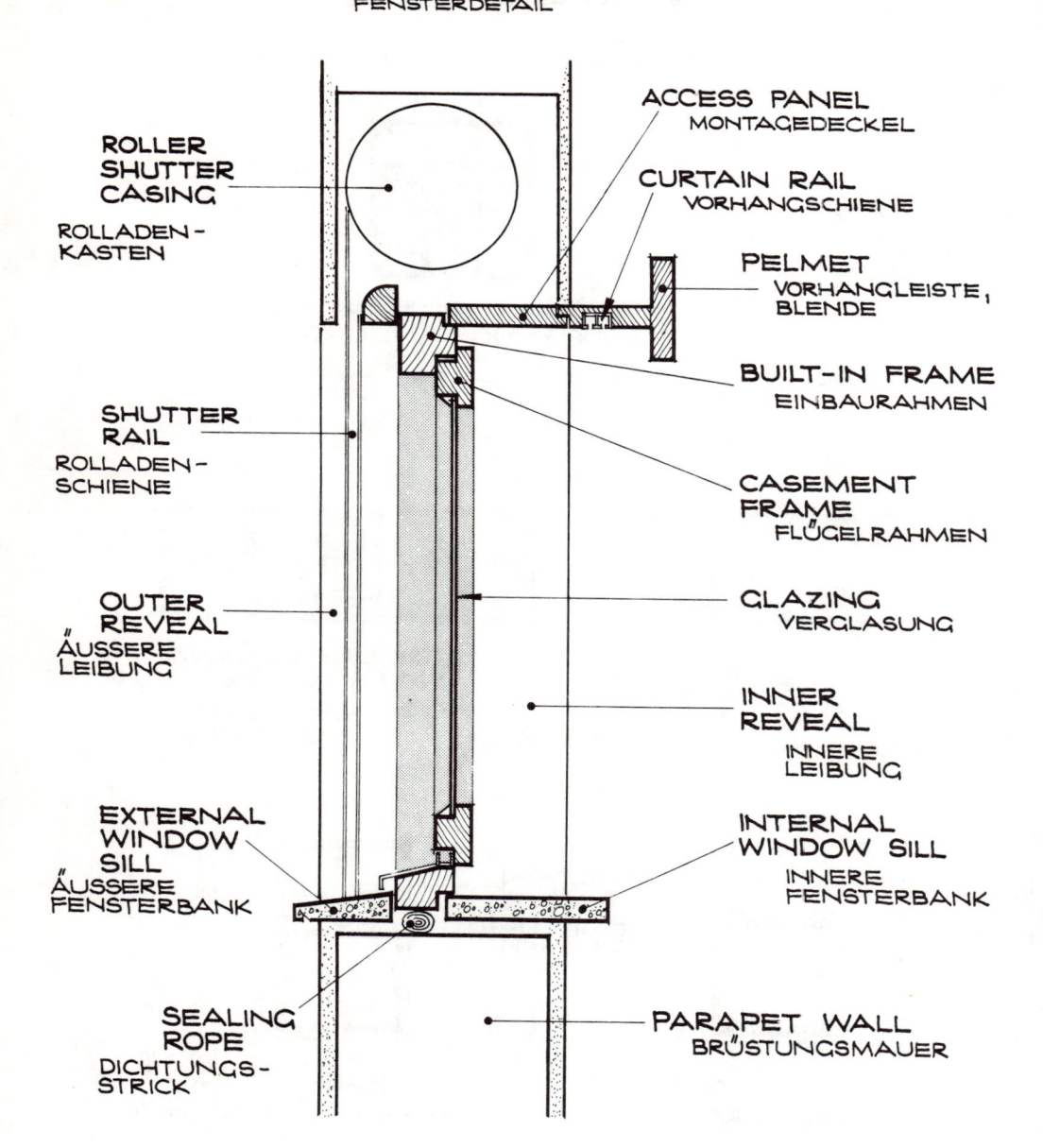

TIMBERWORK
HOLZARBEITEN

- ROLLER SHUTTER CASING — ROLLADENKASTEN
- ACCESS PANEL — MONTAGEDECKEL
- CURTAIN RAIL — VORHANGSCHIENE
- PELMET — VORHANGLEISTE, BLENDE
- BUILT-IN FRAME — EINBAURAHMEN
- SHUTTER RAIL — ROLLADENSCHIENE
- CASEMENT FRAME — FLÜGELRAHMEN
- OUTER REVEAL — ÄUSSERE LEIBUNG
- GLAZING — VERGLASUNG
- INNER REVEAL — INNERE LEIBUNG
- EXTERNAL WINDOW SILL — ÄUSSERE FENSTERBANK
- INTERNAL WINDOW SILL — INNERE FENSTERBANK
- SEALING ROPE — DICHTUNGSSTRICK
- PARAPET WALL — BRÜSTUNGSMAUER

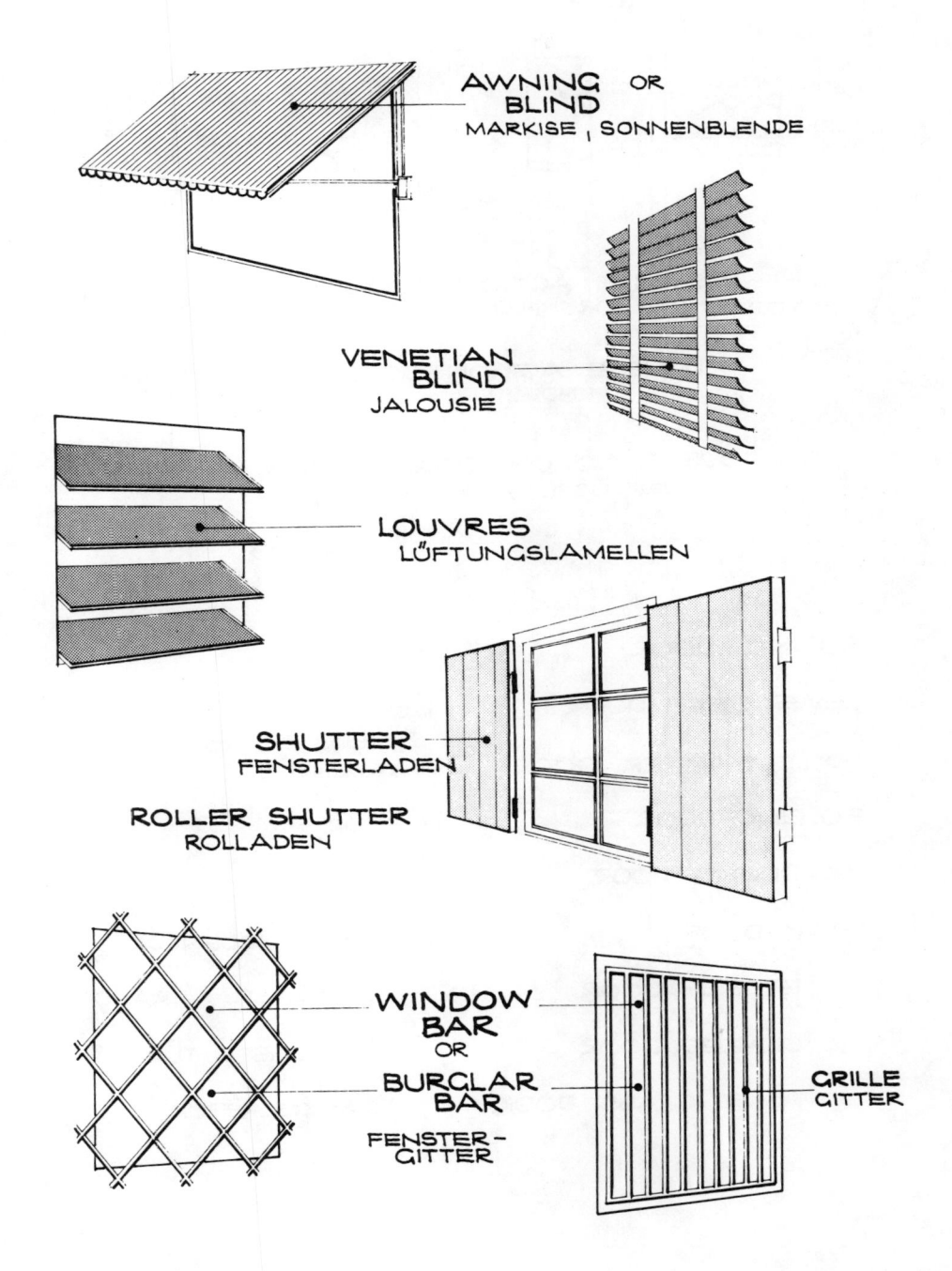

SPECIAL WINDOW FITTINGS
SONDERBAUTEILE FÜR FENSTER

TIMBERWORK
HOLZARBEITEN

AWNING or BLIND
MARKISE, SONNENBLENDE

VENETIAN BLIND
JALOUSIE

LOUVRES
LÜFTUNGSLAMELLEN

SHUTTER
FENSTERLADEN

ROLLER SHUTTER
ROLLADEN

WINDOW BAR
OR
BURGLAR BAR
FENSTERGITTER

GRILLE
GITTER

DOORS
TÜREN

TIMBERWORK
HOLZARBEITEN

INTERNAL DOOR
INNENTÜR

EXTERNAL DOOR
AUSSENTÜR

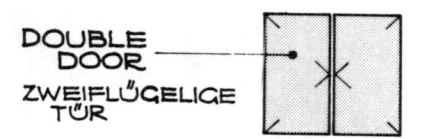

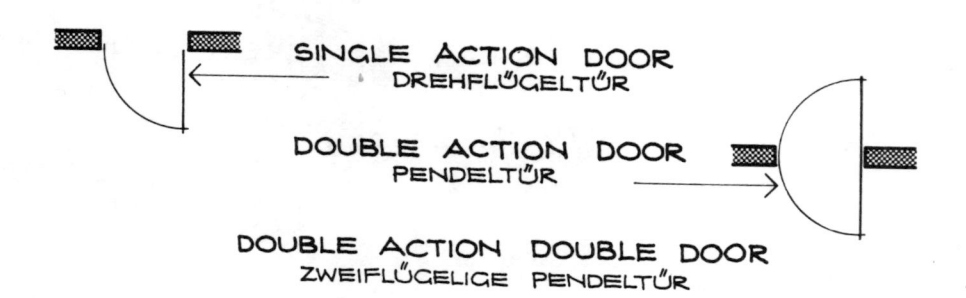

SINGLE DOOR
EINFLÜGELIGE TÜR

DOUBLE DOOR
ZWEIFLÜGELIGE TÜR

SINGLE ACTION DOOR
DREHFLÜGELTÜR

DOUBLE ACTION DOOR
PENDELTÜR

DOUBLE ACTION DOUBLE DOOR
ZWEIFLÜGELIGE PENDELTÜR

SLIDING DOOR	SCHIEBETÜR
LEVER GEAR DOOR	HEBETÜR
ROLLERSHUTTER DOOR	ROLLADENTÜR
FOLDING DOOR	FALTTÜR
REVOLVING DOOR	DREHTÜR
TRAP DOOR	FALLTÜR , BODENTÜR
FRENCH DOOR	VERGLASTE VERANDATÜR
GLAZED DOOR	VERGLASTE TÜR
PLATE GLASS DOOR	GANZGLASTÜR

DOOR FRAMES
TÜRRAHMEN

TIMBERWORK
HOLZARBEITEN

METAL LINING
METALLZARGE

METAL FRAME OF ANGLE OR Z-SECTION
METALLRAHMEN AUS WINKEL- ODER Z-PROFILEN

REVEAL
LEIBUNG

FRAME
STOCK, RAHMEN

DOOR LEAF
TÜRBLATT

CORNER STRIP, BEAD
PUTZLEISTE

BEAD
PUTZLEISTE

LINING
FUTTER

DOOR LEAF
TÜRBLATT

ARCHITRAVE
VERKLEIDUNG

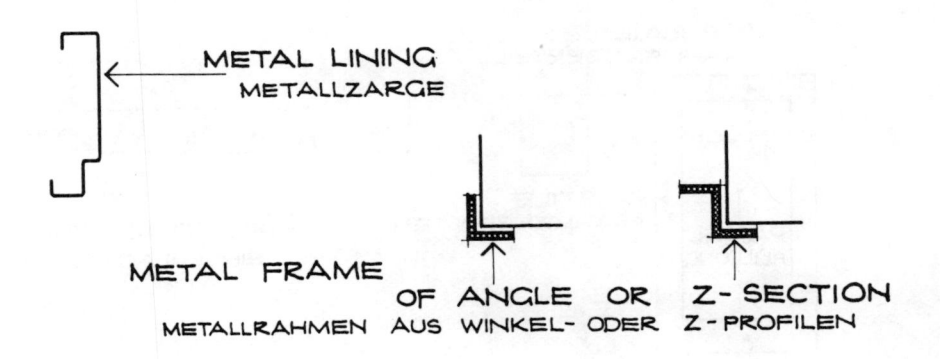

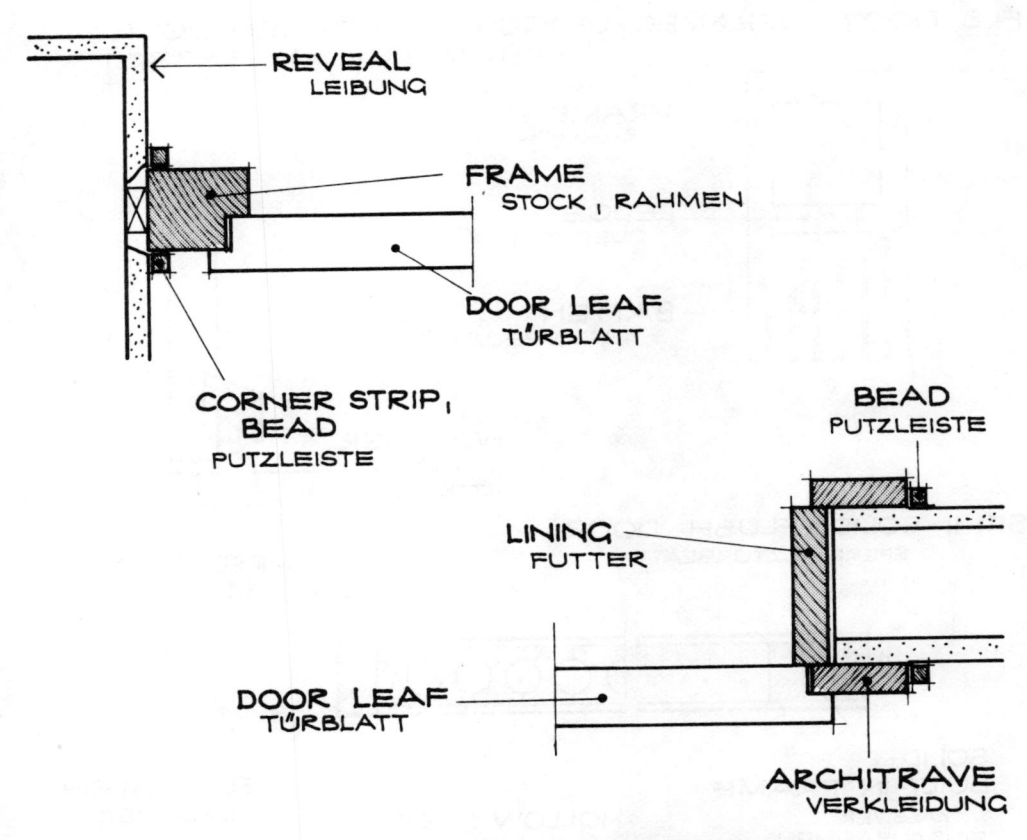

LINING AND ARCHITRAVE
FUTTER UND VERKLEIDUNG

DOOR LEAVES / TÜRBLÄTTER

TIMBERWORK / HOLZARBEITEN

FRAMED DOOR: / RAHMENTÜRE:

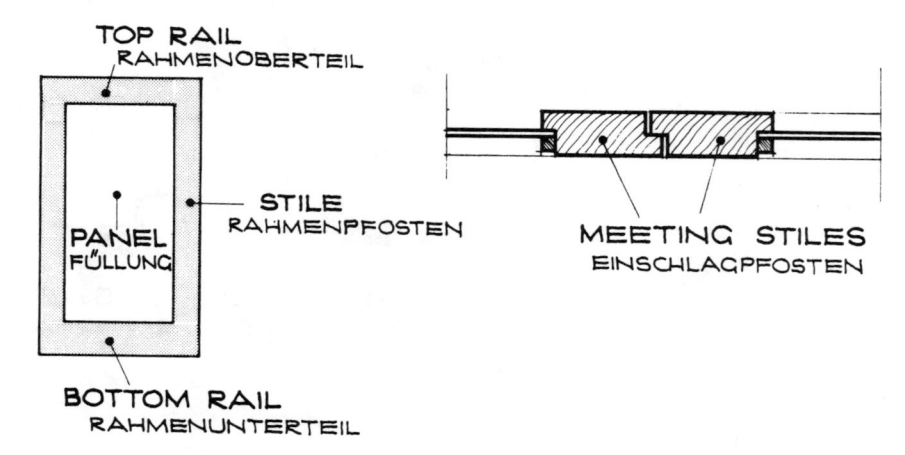

- TOP RAIL / RAHMENOBERTEIL
- STILE / RAHMENPFOSTEN
- PANEL / FÜLLUNG
- BOTTOM RAIL / RAHMENUNTERTEIL
- MEETING STILES / EINSCHLAGPFOSTEN

FLB DOOR = FRAMED, LEDGED & BATTENED DOOR: / AUFGEDOPPELTE RAHMENTÜRE:

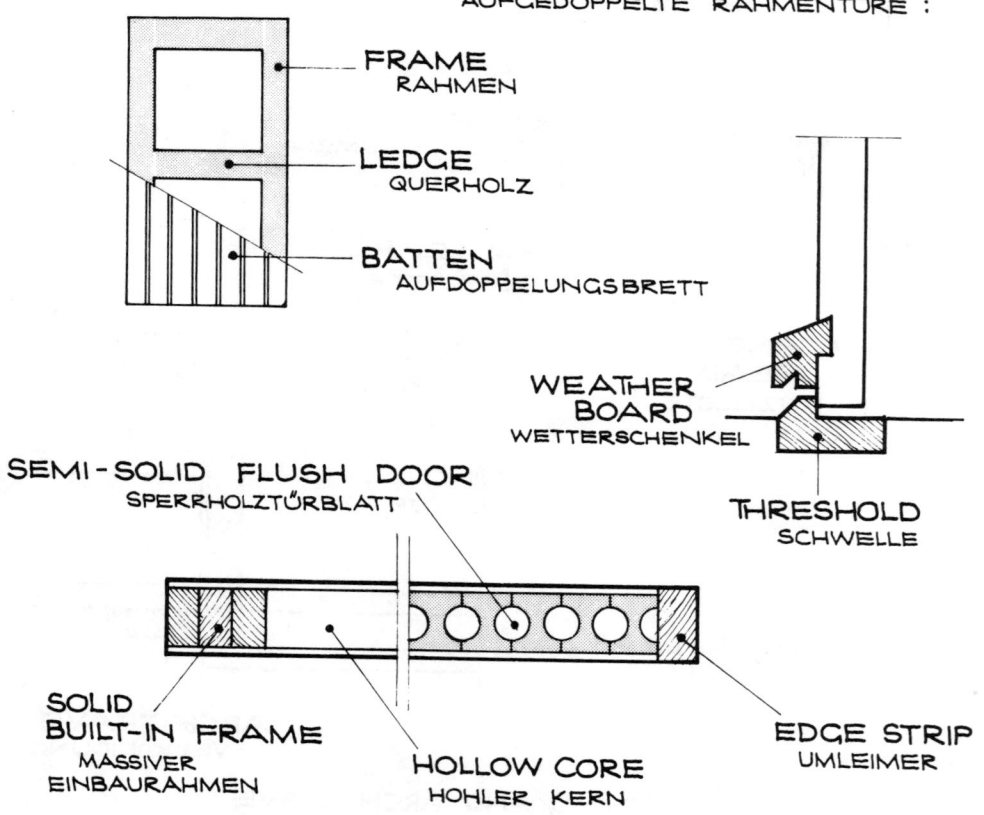

- FRAME / RAHMEN
- LEDGE / QUERHOLZ
- BATTEN / AUFDOPPELUNGSBRETT
- WEATHER BOARD / WETTERSCHENKEL
- THRESHOLD / SCHWELLE

SEMI-SOLID FLUSH DOOR / SPERRHOLZTÜRBLATT

- SOLID BUILT-IN FRAME / MASSIVER EINBAURAHMEN
- HOLLOW CORE / HOHLER KERN
- EDGE STRIP / UMLEIMER

GATES
TORE

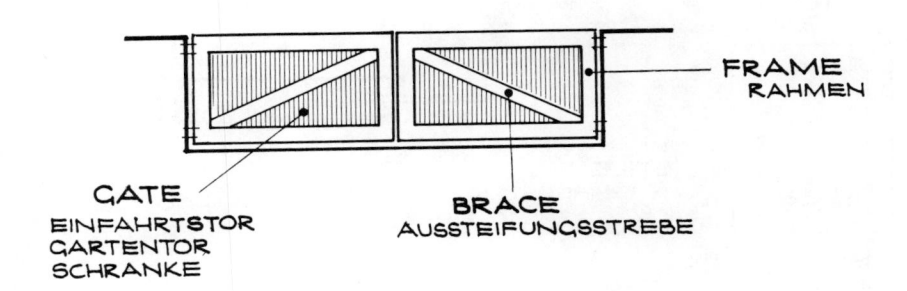

TIMBERWORK
HOLZARBEITEN

FRAME
RAHMEN

GATE
EINFAHRTSTOR
GARTENTOR
SCHRANKE

BRACE
AUSSTEIFUNGSSTREBE

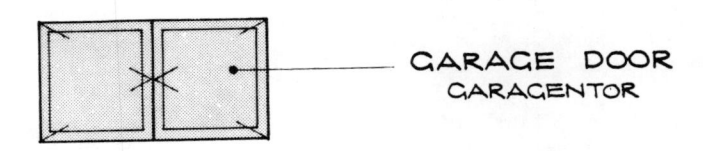

GARAGE DOOR
GARAGENTOR

UP AND OVER DOOR
TIP-UP DOOR,
TILT-UP DOOR
KIPPTOR,
SCHWINGTOR

ROOFS, PLUMBING & FINISHES
DÄCHER, INSTALLATIONEN UND AUSBAU

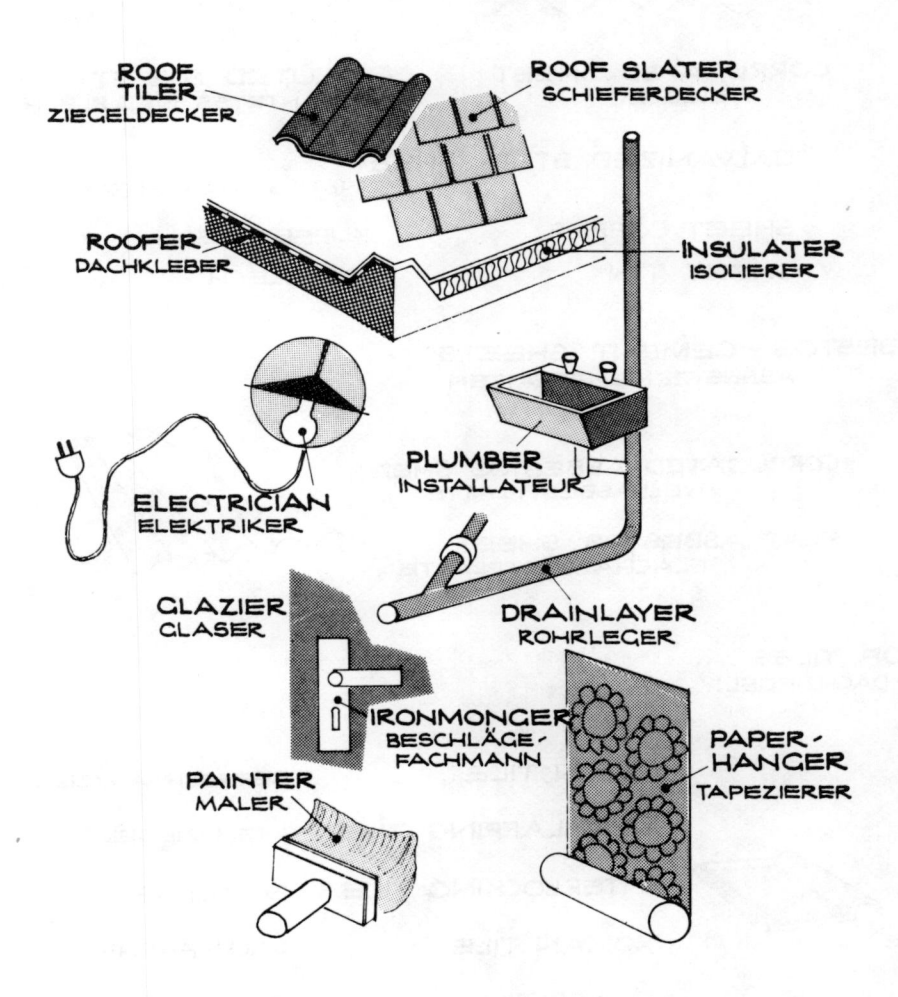

ROOF TILER — ZIEGELDECKER

ROOF SLATER — SCHIEFERDECKER

ROOFER — DACHKLEBER

INSULATER — ISOLIERER

ELECTRICIAN — ELEKTRIKER

PLUMBER — INSTALLATEUR

GLAZIER — GLASER

DRAINLAYER — ROHRLEGER

IRONMONGER — BESCHLÄGEFACHMANN

PAINTER — MALER

PAPERHANGER — TAPEZIERER

ROOF SHEETS & ROOF TILES
DACHPLATTEN UND DACHZIEGEL

ROOF TILER
ZIEGELDECKER

METAL SHEETS:
BLECHE:

CORRUGATED SHEET	FLUTED SHEET
WELLBLECH	GEKANTETES WELLBLECH

GALVANIZED STEEL SHEET — VERZINKTES STAHLBLECH
SHEET COPPER — KUPFERBLECH
SHEET TIN — ZINNBLECH

ASBESTOS - CEMENT SHEETS:
ASBESTZEMENTPLATTEN:

CORRUGATED ASBESTOS SHEET — WELLASBESTPLATTE
FLAT ASBESTOS SHEET — FLACHASBESTPLATTE

ROOF TILES:
DACHZIEGEL:

PLAIN TILE	BIBERSCHWANZZIEGEL
OVERLAPPING TILE	KREMPZIEGEL
INTERLOCKING TILE	FALZZIEGEL
ROMAN TILE	DACHPFANNE
GLASS TILE	GLASDACHZIEGEL
FIBREGLASS TILE	GLASFASERDACHZIEGEL

ROOF SLATES & SHINGLES
DACHSCHIEFER UND SCHINDELN

ROOF SLATER
SCHIEFERDECKER

ROOF SLATE :
DACHSCHIEFER:

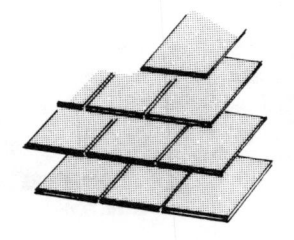

    NATURAL SLATE
        NATURSCHIEFER

    ASBESTOS-CEMENT SLATE
        ASBESTZEMENTSCHIEFER

SHINGLE :
DACHSCHINDEL:

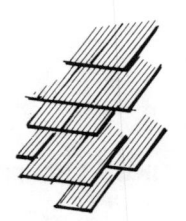

    WOOD SHINGLE
        HOLZSCHINDEL

    STONE SHINGLE
        STEINSCHINDEL

    COPPER SHINGLE
        KUPFERSCHINDEL

UNDERLAY FOIL
    UNTERLEGFOLIE

THATCHED ROOF :
STROHDACH :

GRASS THATCH	GRASDACH
STRAW THATCH	STROHDACH
REED THATCH	RETH- / REET- / REITH- } DACH

BUILT-UP FLAT ROOF
FLACHDACHKONSTRUKTION

ROOFER
DACHDECKER

ROOF COVERING — DACHHAUT

WATER PROOFING — FEUCHTIGKEITSISOLIERUNG

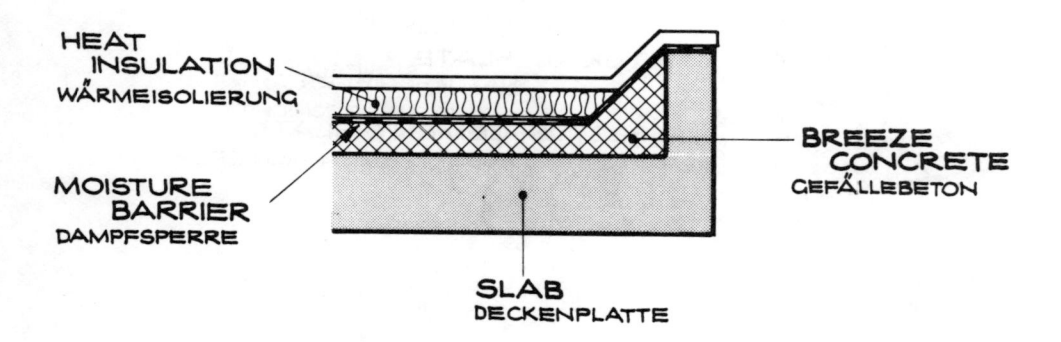

HEAT INSULATION — WÄRMEISOLIERUNG

MOISTURE BARRIER — DAMPFSPERRE

BREEZE CONCRETE — GEFÄLLEBETON

SLAB — DECKENPLATTE

EXPANSION JOINT SEALING — DEHNFUGENABDICHTUNG

FIBREGLASS DOME LIGHT — GLASFASERLICHTKUPPEL

ROOF LIGHT FLASHING — OBERLICHTANSCHLUSSBLECH

ROOF OUTLET FLASHING — DACHABLAUFEINFASSBLECH

MOISTURE BARRIERS: — DAMPFSPERREN:

ALUMINIUM FOIL — ALUMINIUMFOLIE

PVC SHEET — PVC-BAHN

POLYTHENE SHEET — POLYÄTHYLENFOLIE

POLYURETHANE SHEET — KUNSTSTOFF-FOLIE

PLASTIC WELDING — PLASTIKSCHWEISSEN

WATER PROOFING
FEUCHTIGKEITSISOLIERUNG

INSULATER
ISOLIERER

ROOFING FELT :
  DACHPAPPE :

- BITUMINOUS FELT — BITUMENPAPPE
  - FIVE - PLY — FÜNFSCHICHTIG
  - MULTI - PLY — MEHRSCHICHTIG
- IMPREGNATED FELT — IMPRÄGNIERTE PAPPE
- PVC SHEET — PVC - BAHN

BONDING AGENT
  BINDEMITTEL :

- MOLTEN BITUMEN — HEISSBITUMEN
- MASTIC ASPHALT — WEICHASPHALT

HOT OR COLD APPLICATION — HEISS- ODER KALTAUFTRAG
BRUSH OR SPRAY APPLICATION — BÜRST- ODER SPRÜHAUFTRAG

MEMBRANE :
  ISOLIERHAUT :

- PLASTIC MEMBRANE — PLASTIKHAUT
- TAR MEMBRANE — TEERISOLIERHAUT
- EMULSION MEMBRANE — EMULSIONSHAUT
- EPOXY MEMBRANE — EPOXYHAUT

TANKING
WANNENAUSBILDUNG

HEAT INSULATION
WÄRME ISOLIERUNG

INSULATER / ISOLIERER

INSULATING MATERIALS COME USUALLY IN
ISOLIERMATERIALIEN GIBT ES GEWÖHNLICH IN FORM VON

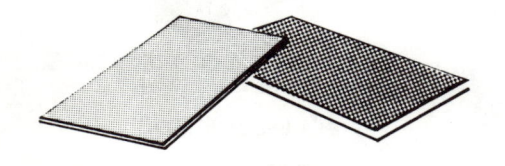

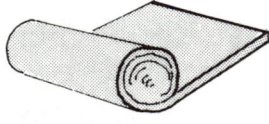

SHEETS	BOARDS	ROLLS	OR	FLAKES
PLATTEN (DÜNN)	TAFELN (DICK)	BAHNEN	ODER	FLOCKEN

POLYSTYRENE SHEET, BOARD OR FLAKES
STYROPOR-PLATTE, TAFEL ODER FLOCKEN

FIBREGLASS SHEET — GLASFASERPLATTE

ASBESTOS-CEMENT SHEET — ASBESTZEMENTPLATTE

PEAT MULL — TORFMULL

EXPANDED CORK SHEET — EXPANDIERTE KORKPLATTE

GLASS WOOL ROLL — GLASWOLLBAHN

WOOD WOOL — HOLZWOLLE

MINERAL WOOL — MINERALWOLLE

ROCK WOOL — STEINWOLLE

SLAG WOOL — SCHLACKENWOLLE

THERMAL FOIL — WÄRMEISOLIERENDE FOLIE

SEALING ROPE — DICHTUNGSSTRICK

INSULATING BRICK — ISOLIERHOHLBLOCK

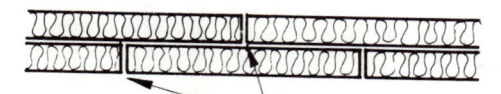

STAGGERED JOINTS
VERSETZTE FUGEN

ELECTRICAL WORK
ELEKTROARBEITEN

ELECTRICIAN
ELEKTRIKER

CONNECTION TO MAIN SUPPLY
ANSCHLUSS ZUM HAUPTNETZ

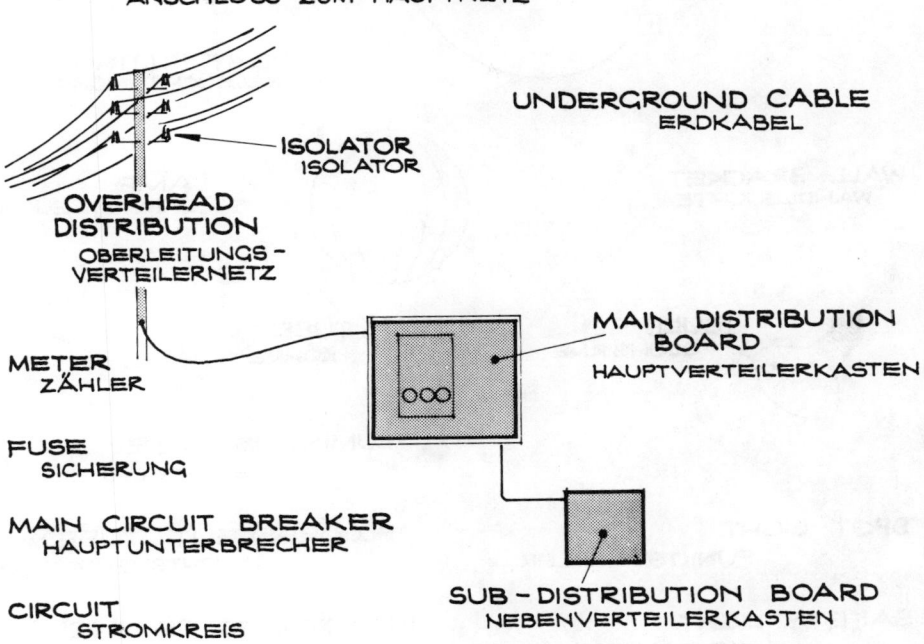

ISOLATOR
ISOLATOR

OVERHEAD DISTRIBUTION
OBERLEITUNGS-VERTEILERNETZ

UNDERGROUND CABLE
ERDKABEL

MAIN DISTRIBUTION BOARD
HAUPTVERTEILERKASTEN

METER
ZÄHLER

FUSE
SICHERUNG

MAIN CIRCUIT BREAKER
HAUPTUNTERBRECHER

SUB-DISTRIBUTION BOARD
NEBENVERTEILERKASTEN

CIRCUIT
STROMKREIS

CONDUCTORS:
STROMLEITER

    BARE CONDUCTOR — WIRE
      NICHTISOLIERTER LEITER   DRAHT

    INSULATED CONDUCTOR — CABLE
      ISOLIERTER LEITER   KABEL

TENSION
  STROMSPANNUNG

HIGH TENSION
  HOCHSPANNUNG

POWER CURRENT
  STARKSTROM

SINGLE PHASE
  EINPHASIG

THREE PHASE CURRENT
  DREIPHASENSTROM

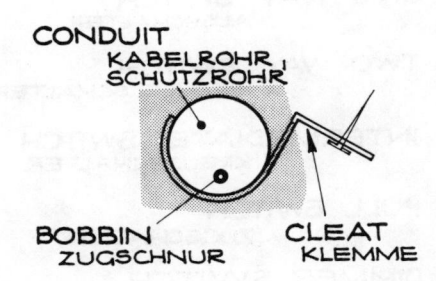

EARTHING
  ERDUNG

CONDUIT
KABELROHR, SCHUTZROHR

BOBBIN
ZUGSCHNUR

CLEAT
KLEMME

CHASING & MAKING GOOD
SCHLITZE STEMMEN UND EINPUTZEN

LIGHTING
BELEUCHTUNG

ELECTRICIAN
ELEKTRIKER

LIGHT POINT
LICHTANSCHLUSS-STELLE

(LIGHTNING)
(BLITZ)

LIGHT FITTING
BELEUCHTUNGSKÖRPER

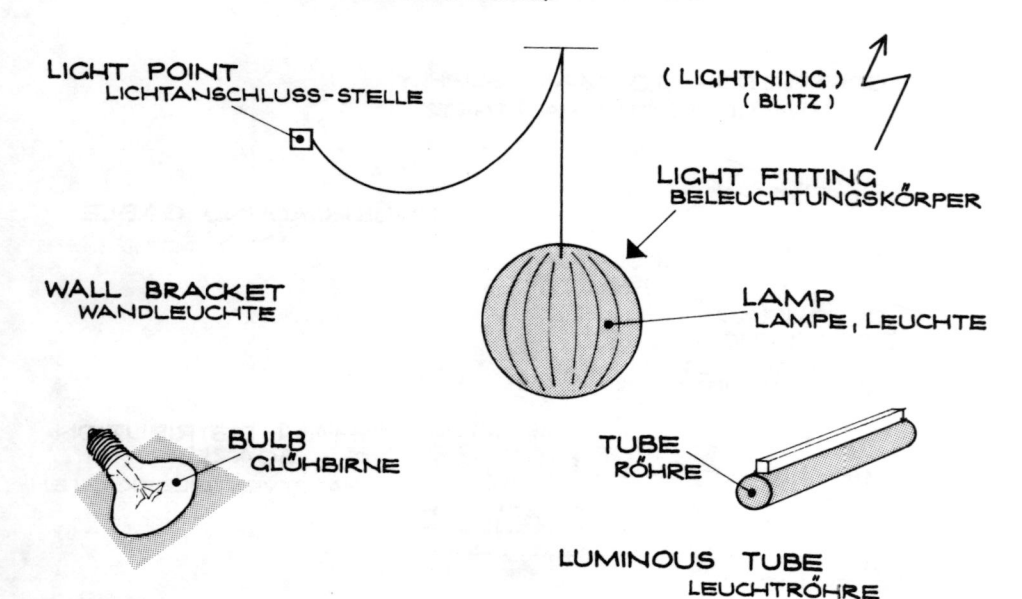

WALL BRACKET
WANDLEUCHTE

LAMP
LAMPE, LEUCHTE

BULB
GLÜHBIRNE

TUBE
RÖHRE

LUMINOUS TUBE
LEUCHTRÖHRE

SPOT LIGHT
PUNKTSTRAHLER

FLUORESCENT TUBE
LEUCHTSTOFFRÖHRE

SAFETY LIGHT
SICHERHEITSLEUCHTE

ILLUMINATED CEILING
LEUCHTDECKE

SWITCHES
SCHALTER

ONE WAY SWITCH
AUSSCHALTER

TWO WAY SWITCH
WECHSELSCHALTER

INTERMEDIATE SWITCH
KREUZSCHALTER

PULL SWITCH
ZUGSCHALTER

DIMMER SWITCH
DÄMMERUNGSSCHALTER,
DIMMER

COVER PLATE
TO SWITCH
SCHALTERDECKEL

PLUGS
STECKDOSEN

ELECTRICIAN
ELEKTRIKER

POWER POINT
  KRAFTANSCHLUSS-STELLE

SOCKET
  STECKDOSENGEHÄUSE

POWER SKIRTING
  ELEKTROSOCKELLEISTE

POWER POINTS (PLUGS) TO
  STECKDOSENANSCHLÜSSE FÜR

GEYSER	WARMWASSERGERÄT
STOVE	HERD
HEATER	HEIZGERÄT
CLOCK	UHRWERK
MOTOR	ELEKTROMOTOR
FAN ETC	VENTILATOR ETC

LOW VOLTAGE FITTINGS
SCHWACHSTROMARMATUREN

SWITCHBOARD AND TELEPHONE
  VERMITTLUNGSAPPARAT UND TELEFON

INTERCOM AND SPEAKER
  LAUTSPRECHERANLAGE UND LAUTSPRECHER

BELL AND BELL PUSH
  KLINGEL UND KLINGELKNOPF

BUZZER	SUMMER
AERIAL	ANTENNE
THERMOSTAT	THERMOSTAT

WATER SUPPLY
WASSERVERSORGUNG

PLUMBER
INSTALLATEUR

COLD WATER SUPPLY
KALTWASSERVERSORGUNG

FRESH WATER
SÜSSWASSER

SALT WATER
SALZWASSER

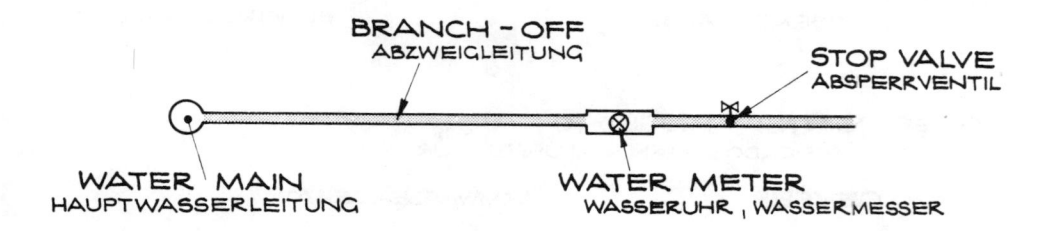

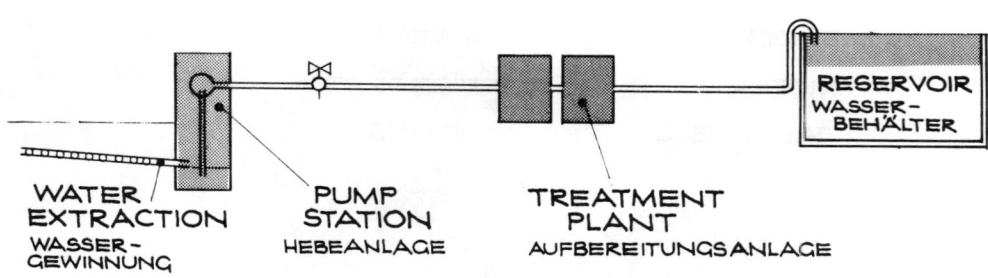

BRANCH - OFF ABZWEIGLEITUNG

STOP VALVE ABSPERRVENTIL

WATER MAIN HAUPTWASSERLEITUNG

WATER METER WASSERUHR, WASSERMESSER

WATER EXTRACTION WASSERGEWINNUNG

PUMP STATION HEBEANLAGE

TREATMENT PLANT AUFBEREITUNGSANLAGE

RESERVOIR WASSERBEHÄLTER

IRRIGATION SYSTEM BEWÄSSERUNGSANLAGE

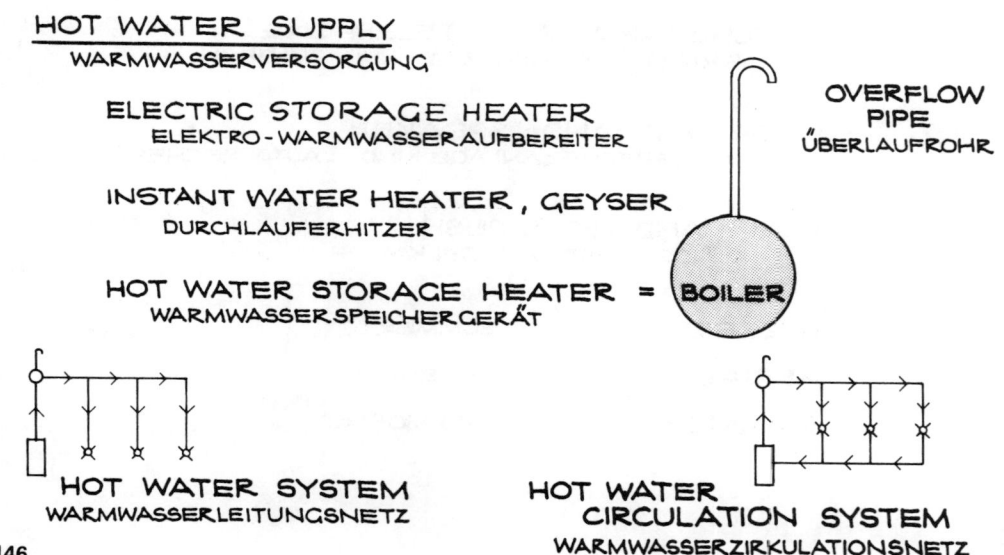

HOT WATER SUPPLY
WARMWASSERVERSORGUNG

ELECTRIC STORAGE HEATER
ELEKTRO - WARMWASSERAUFBEREITER

INSTANT WATER HEATER, GEYSER
DURCHLAUFERHITZER

HOT WATER STORAGE HEATER = BOILER
WARMWASSERSPEICHERGERÄT

OVERFLOW PIPE ÜBERLAUFROHR

HOT WATER SYSTEM WARMWASSERLEITUNGSNETZ

HOT WATER CIRCULATION SYSTEM WARMWASSERZIRKULATIONSNETZ

PLUMBER'S & DRAINLAYER'S TOOLS — PLUMBER
INSTALLATEURSWERKZEUGE — INSTALLATEUR

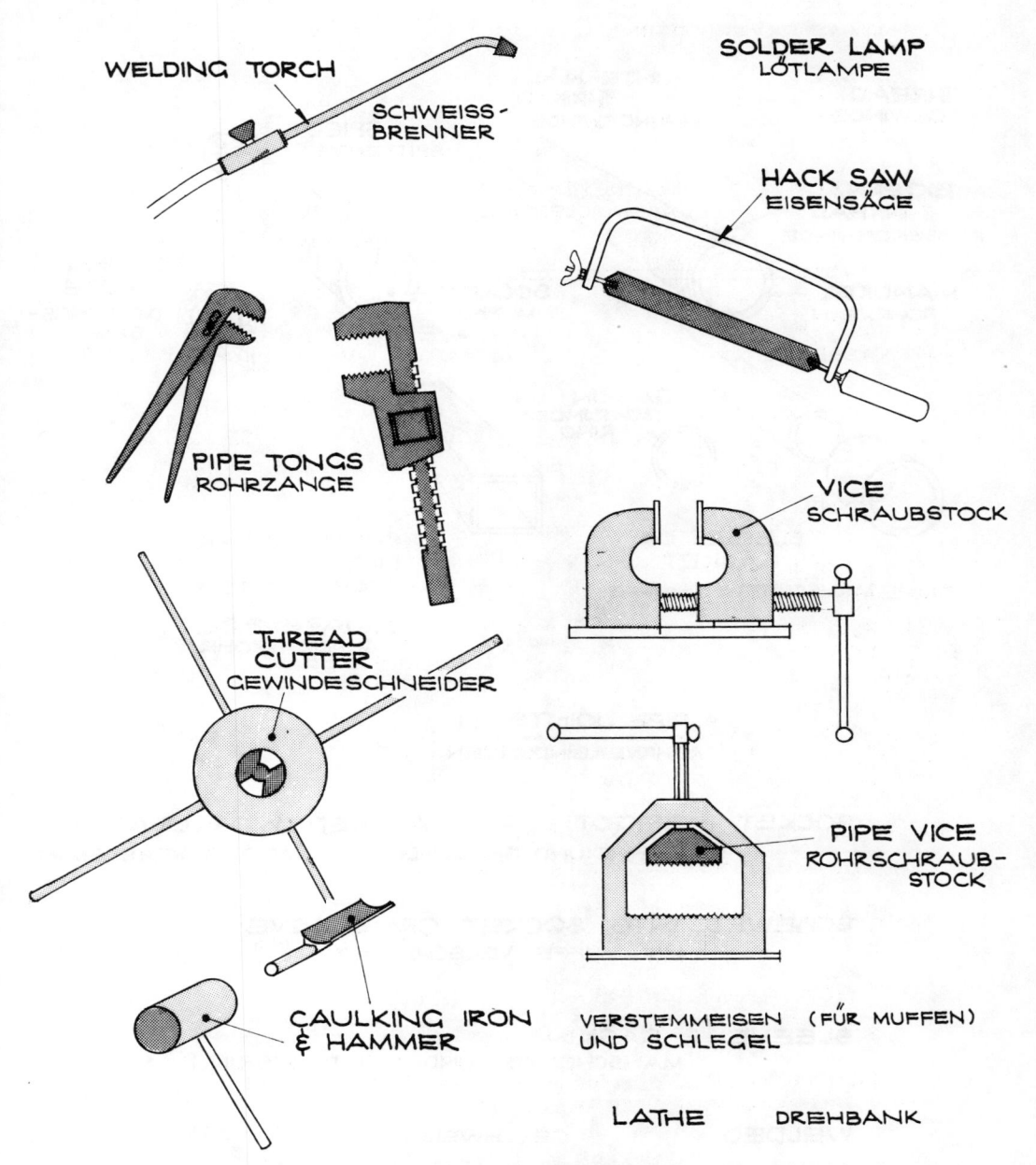

WELDING TORCH — SCHWEISSBRENNER

SOLDER LAMP — LÖTLAMPE

HACK SAW — EISENSÄGE

PIPE TONGS — ROHRZANGE

VICE — SCHRAUBSTOCK

THREAD CUTTER — GEWINDESCHNEIDER

PIPE VICE — ROHRSCHRAUBSTOCK

CAULKING IRON & HAMMER — VERSTEMMEISEN (FÜR MUFFEN) UND SCHLEGEL

LATHE — DREHBANK

PIPE BENDING MACHINE — ROHRBIEGEMASCHINE

PIPES
LEITUNGSROHRE

DRAINLAYER & PLUMBER
INSTALLATEUR

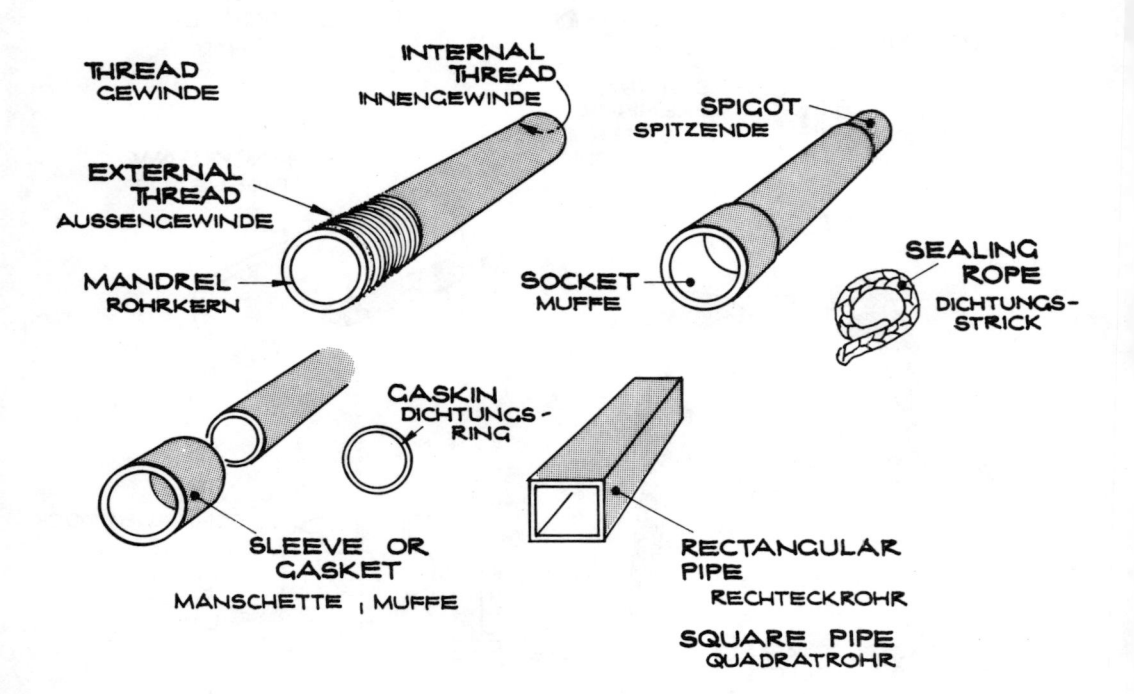

- **THREAD** — GEWINDE
- **INTERNAL THREAD** — INNENGEWINDE
- **EXTERNAL THREAD** — AUSSENGEWINDE
- **MANDREL** — ROHRKERN
- **SPIGOT** — SPITZENDE
- **SOCKET** — MUFFE
- **SEALING ROPE** — DICHTUNGSSTRICK
- **GASKIN** — DICHTUNGSRING
- **SLEEVE OR GASKET** — MANSCHETTE, MUFFE
- **RECTANGULAR PIPE** — RECHTECKROHR
- **SQUARE PIPE** — QUADRATROHR

PIPE JOINTS
ROHRVERBINDUNGEN

SOCKET & SPIGOT — CAULKED WITH LEAD
MUFFE UND SPITZENDE — MIT BLEI VERSTEMMT

SCREWED INTO SOCKET OR SLEEVE
MIT MUFFE VERSCHRAUBT

SLEEVE & GASKIN
MANSCHETTE UND DICHTUNGSRING

WELDED   GESCHWEISST

SPECIAL COUPLING   SPEZIALROHRKUPPLUNG

PIPE FITTINGS
ROHRFORMSTÜCKE

DRAINLAYER & PLUMBER
INSTALLATEUR

BEND — BOGEN

OFFSET — ABSATZSTÜCK

SHOE — AUSLAUFSTÜCK

SINGLE BRANCH — EINFACH-ABZWEIG

DOUBLE BRANCH — DOPPELABZWEIG

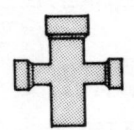

BOSS BRANCH — DOPPEL T-STÜCK

REDUCER — REDUZIERSTÜCK

INSPECTION EYE — REINIGUNGSSTÜCK

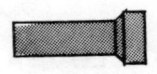

CONNECTOR — ANSCHLUSS-STÜCK

PIPE MATERIALS
ROHRMATERIALIEN

STEEL TUBE	STAHLROHR
CAST IRON PIPE	GUSSROHR
GALVANIZED STEEL PIPE	VERZINKTES STAHLROHR
COPPER PIPE	KUPFERROHR
PVC PIPE	PVC-ROHR
STONEWARE PIPE / GLAZED EARTHENWARE PIPE	STEINZEUGROHR, TONROHR
CONCRETE PIPE	BETONROHR
ASBESTOS-CEMENT PIPE	ASBESTZEMENTROHR

WATER SUPPLY FITTINGS
WASSER - ARMATUREN

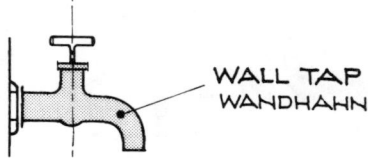

PLUMBER
INSTALLATEUR

TAP
WASSERHAHN

WALL TAP — WANDHAHN

STOPCOCK — ABSPERRHAHN
STOPVALVE — ABSPERRVENTIL

PILLAR-TYPE BASIN TAP — STANDHAHN

HOSE CONNECTION — SCHLAUCHANSCHLUSS

STAND PIPE — STANDROHR

MIXER
MISCHBATTERIE

LEVER — UMLEGHEBEL
COVER PLATE — ROSETTE

LEVER ACTION MIXER WITH ANGLE VALVES
MISCHBATTERIE MIT UMLEGHEBEL UND HÄHNEN MIT SEITLICHEM ABGANG

SHOWER SPRAY — SPRÜHKOPF

WASTE
ABLAUF

WASTE PLUG & CHAIN — ABLAUFPFROPFEN UND KETTE

PLASTIC HOSE — PLASTIKSCHLAUCH

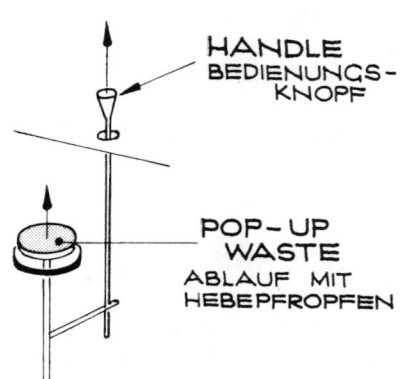

WASTE OUTLET — ABLAUFÖFFNUNG

BOTTLE TRAP — FLASCHENSIPHON

HANDLE — BEDIENUNGSKNOPF

POP-UP WASTE — ABLAUF MIT HEBEPFROPFEN

WASTE GRATING — ABLAUFROST

BASIN & SINK
WASCH - UND SPÜLBECKEN

PLUMBER
INSTALLATEUR

BASIN
WASCHBECKEN

HANDBASIN
HANDWASCHBECKEN

LAVATORY BASIN
WASCHTISCH

MIRROR
SPIEGEL

TUMBLER
TRINKBECHER

TOWEL RAIL
HANDTUCHSTANGE

WASH BASIN
WASCHBECKEN

PLATE GLASS SHELF
GLASABLAGE

BOTTLE TRAP
FLASCHENSIPHON

CONCEALED BRACKETS
VERDECKTE KONSOLEN

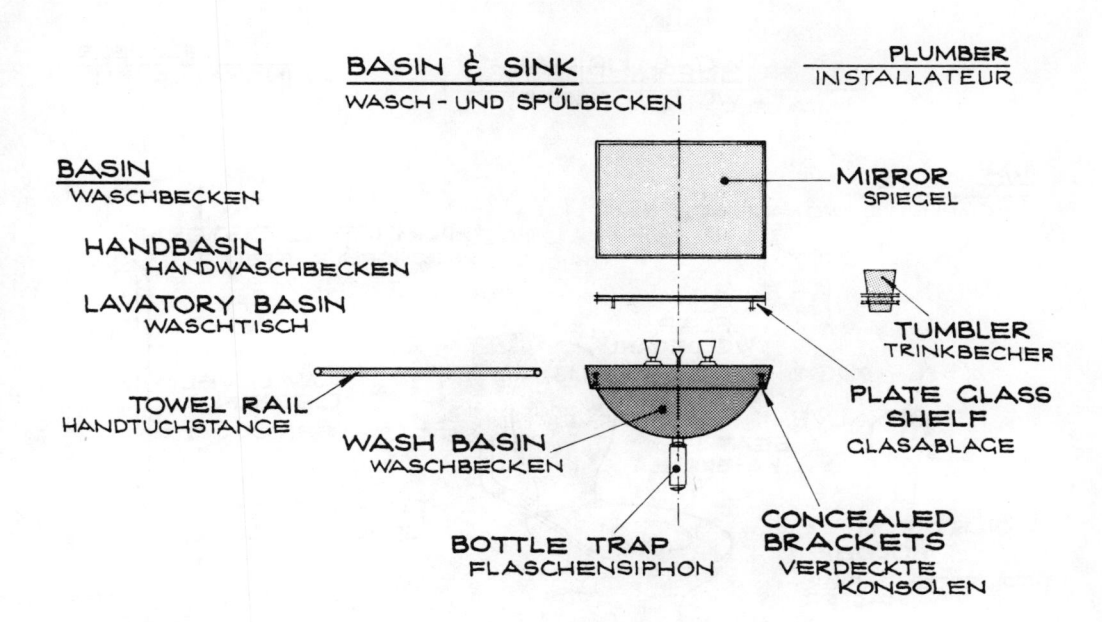

SINK
SPÜLE

CENTRE BOWL SINK
SPÜLE MIT MITTELBECKEN

DOUBLE BOWL SINK
SPÜLE MIT DOPPELBECKEN

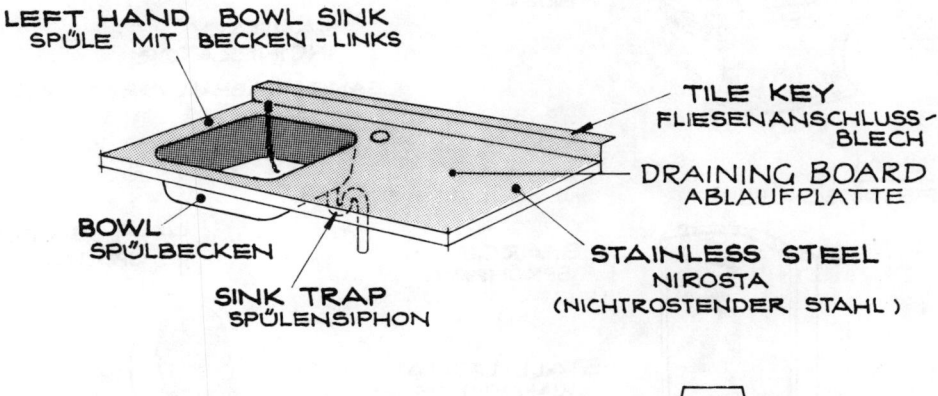

LEFT HAND BOWL SINK
SPÜLE MIT BECKEN - LINKS

TILE KEY
FLIESENANSCHLUSS-BLECH

DRAINING BOARD
ABLAUFPLATTE

BOWL
SPÜLBECKEN

STAINLESS STEEL
NIROSTA
(NICHTROSTENDER STAHL)

SINK TRAP
SPÜLENSIPHON

WATER DISPENSER
FRISCHWASSERGERÄT

HOT WATER DISPENSER
WARMWASSERBEREITER

WC & URINAL
WC UND PISSOIR

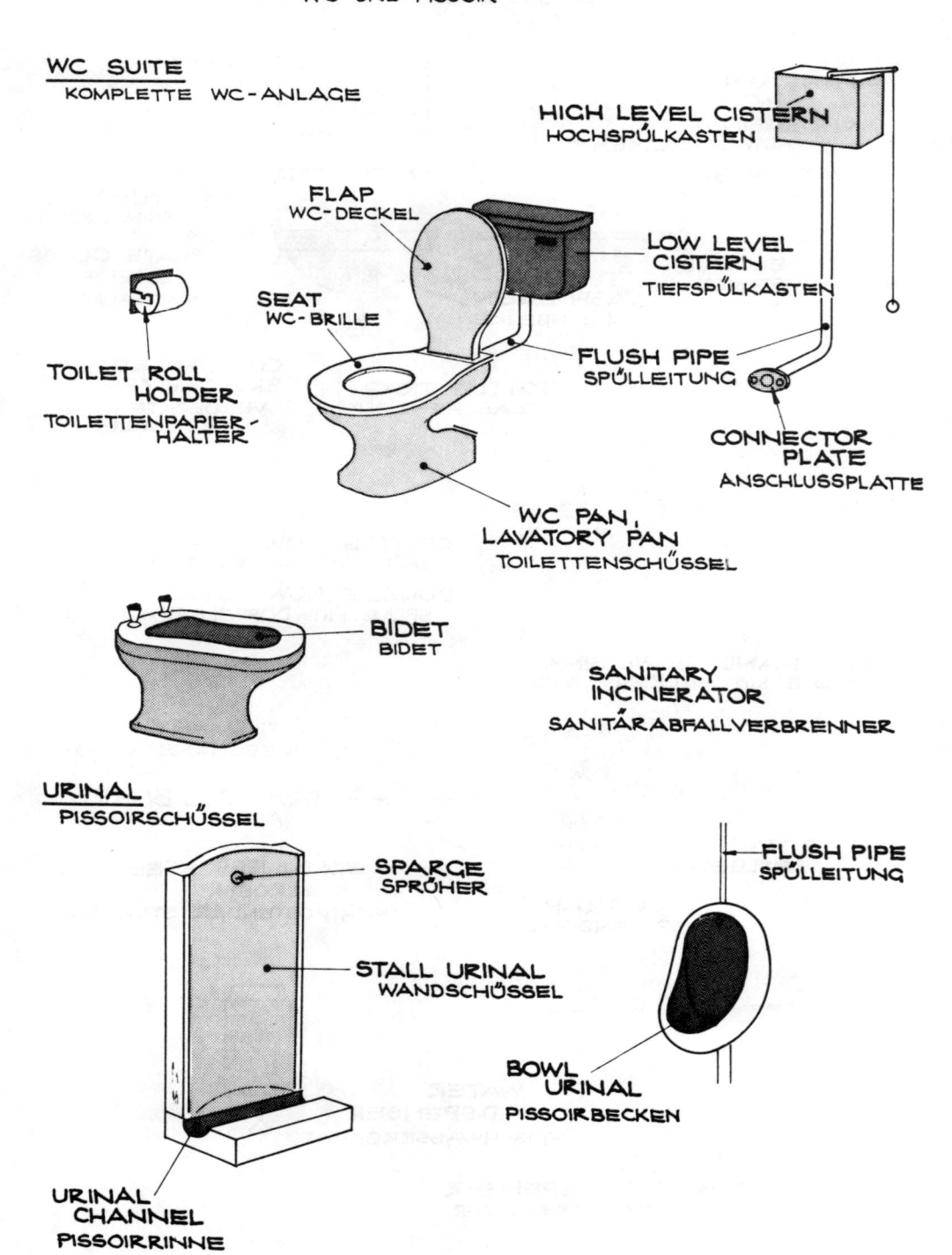

PLUMBER
INSTALLATEUR

WC SUITE
KOMPLETTE WC-ANLAGE

- **HIGH LEVEL CISTERN** — HOCHSPÜLKASTEN
- **FLAP** — WC-DECKEL
- **LOW LEVEL CISTERN** — TIEFSPÜLKASTEN
- **SEAT** — WC-BRILLE
- **FLUSH PIPE** — SPÜLLEITUNG
- **TOILET ROLL HOLDER** — TOILETTENPAPIERHALTER
- **CONNECTOR PLATE** — ANSCHLUSSPLATTE
- **WC PAN, LAVATORY PAN** — TOILETTENSCHÜSSEL

BIDET — BIDET

SANITARY INCINERATOR — SANITÄRABFALLVERBRENNER

URINAL
PISSOIRSCHÜSSEL

- **SPARGE** — SPRÜHER
- **STALL URINAL** — WANDSCHÜSSEL
- **FLUSH PIPE** — SPÜLLEITUNG
- **BOWL URINAL** — PISSOIRBECKEN
- **URINAL CHANNEL** — PISSOIRRINNE

BATH & SHOWER
BAD UND DUSCHE

PLUMBER / INSTALLATEUR

BATH / BAD

- BATH TUB — BADEWANNE
- SOAP DISH — SEIFENSCHÜSSEL
- TOWEL RAIL — HANDTUCHSTANGE
- DWARF WALL — AUSMAUERUNG
- BATH TRAP — BADSIPHON
- ACCESS PANEL — ZUGANGSTÜRCHEN
- BATHROOM CABINET — SPIEGELSCHRANK
- FLUE — ABZUG
- FLOW TYPE CALORIFIER — DURCHLAUF-ERHITZER
- SHAVER PLUG — RASIERSTECKDOSE
- LIGHT FITTING — LEUCHTE

SHOWER / DUSCHE

- SWIVEL ARM — SCHWENKARM
- WATER & GAS CONNECTION — WASSER- UND GASANSCHLUSS
- SHOWER ROSE — BRAUSEKOPF
- THERMOSTAT
- UNDERPLASTER MIXER — UNTERPUTZ-MISCHBATTERIE
- SHOWER CURTAIN — DUSCHE-VORHANG
- SHOWER BASIN — DUSCHEWANNE
- WASTE & SHOWER TRAP — ABLAUF UND DUSCHESIPHON

SEWAGE DISCHARGE
ABWASSERBESEITIGUNG

DRAINLAYER / ROHRLEGER

SEWAGE WATER	ABWASSER
WASTE WATER	BRAUCHWASSER
SOIL WATER	SCHMUTZWASSER

SOIL & VENT STACK — SCHMUTZWASSER- UND ENTLÜFTUNGSSTRANG

SOIL STACK — SCHMUTZWASSERSTRANG

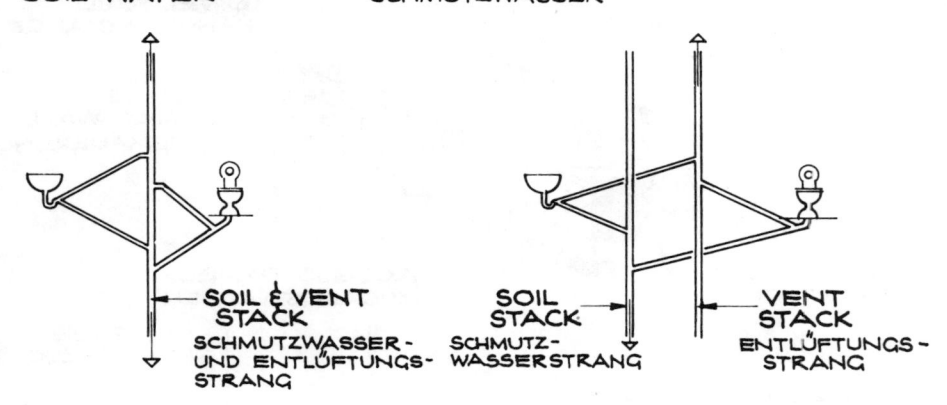

VENT STACK — ENTLÜFTUNGSSTRANG

SINGLE STACK SYSTEM — EINSTRANGSYSTEM

DOUBLE STACK SYSTEM — DOPPELSTRANGSYSTEM

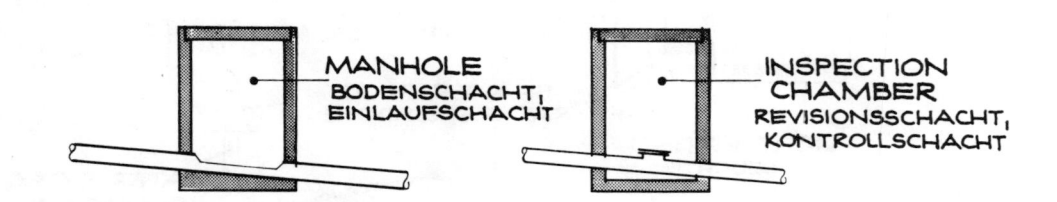

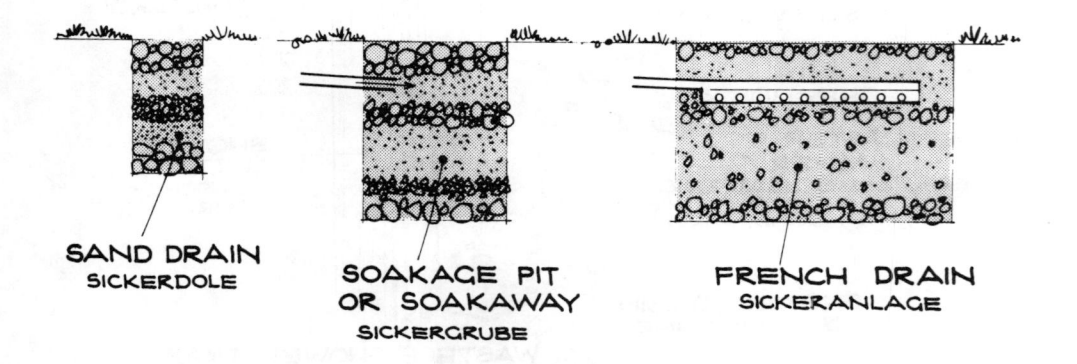

MANHOLE — BODENSCHACHT, EINLAUFSCHACHT

INSPECTION CHAMBER — REVISIONSSCHACHT, KONTROLLSCHACHT

SAND DRAIN — SICKERDOLE

SOAKAGE PIT OR SOAKAWAY — SICKERGRUBE

FRENCH DRAIN — SICKERANLAGE

TRAPS
ABSCHEIDER

DRAINLAYER
ROHRLEGER

SAND TRAP SANDFANG

GREASE TRAP FETTABSCHEIDER

PETROL TRAP,
PETROL INCEPTOR
  BENZINABSCHEIDER

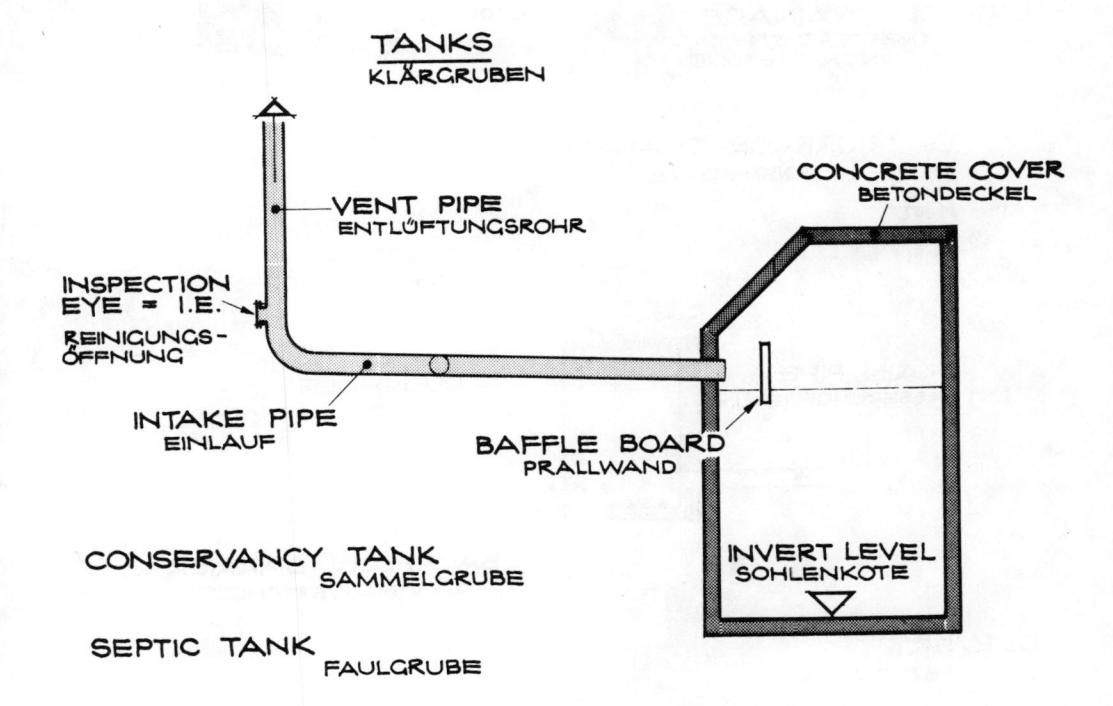

TRAP GERUCHVERSCHLUSS

TANKS
KLÄRGRUBEN

VENT PIPE ENTLÜFTUNGSROHR

INSPECTION EYE = I.E.
REINIGUNGSÖFFNUNG

INTAKE PIPE EINLAUF

BAFFLE BOARD PRALLWAND

CONCRETE COVER BETONDECKEL

INVERT LEVEL SOHLENKOTE

CONSERVANCY TANK SAMMELGRUBE

SEPTIC TANK FAULGRUBE

SLUDGE DIGESTION TANK AUSFAULGRUBE

SETTLING BASIN ABSETZBECKEN

STORMWATER DRAINAGE
OBERFLÄCHENENTWÄSSERUNG

DRAINLAYER
ROHRLEGER

ROOF DRAINAGE
DACHENTWÄSSERUNG

BALLOON GRATING — LAUBSIEB
ROOF OUTLET — DACHABLAUF

FLAT ROOF DRAINAGE
FLACHDACHENTWÄSSERUNG

TERRACE DRAINAGE
TERRASSENENTWÄSSERUNG

BALCONY DRAINAGE
BALKONENTWÄSSERUNG

GRATING — ROST
FLOOR GULLY — BODENABLAUF

SURFACE DRAINAGE
OBERFLÄCHEN-ENTWÄSSERUNG

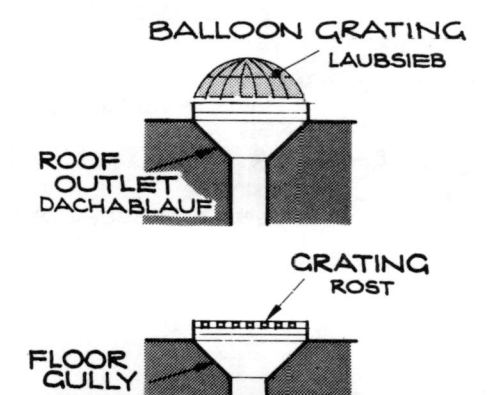

SURFACE CHANNEL
BODENRINNE

GRATING — ABDECKROST

DRAIN PIPE
ENTWÄSSERUNGSLEITUNG

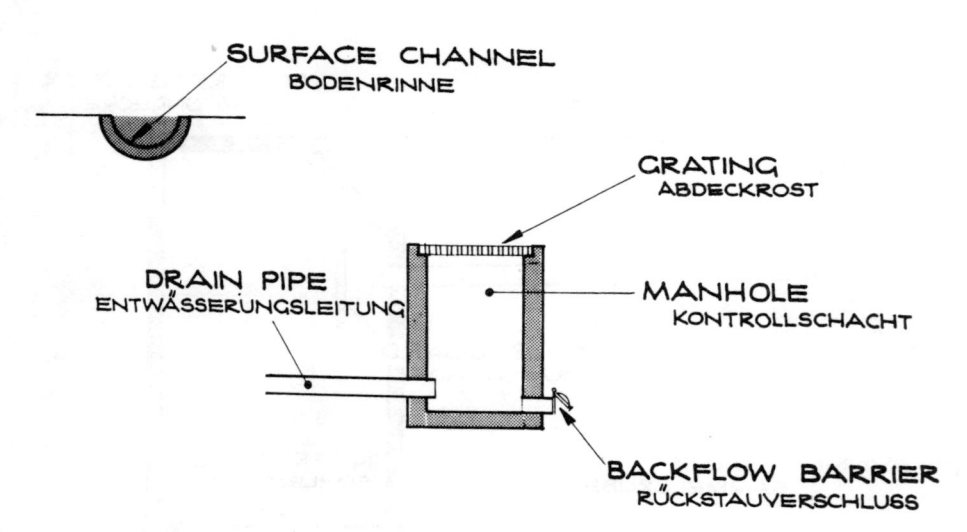

MANHOLE
KONTROLLSCHACHT

BACKFLOW BARRIER
RÜCKSTAUVERSCHLUSS

CULVERT
ABZUGSKANAL

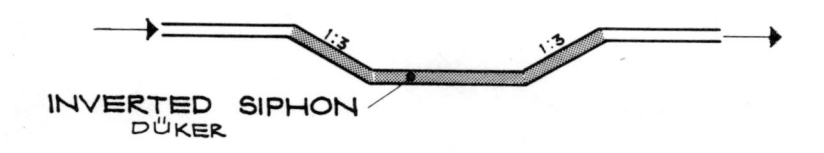

INVERTED SIPHON
DÜKER

GLASS
CLAS

GLAZIER
GLASER

CLEAR GLASS
KLARGLAS

OPAQUE GLASS
UNDURCHSICHTIGES GLAS

TINTED GLASS
FARBGLAS

TRANSLUCENT GLASS
LICHTDURCHLÄSSIGES GLAS

SHEET GLASS — TAFELGLAS

PLATE GLASS — KRISTALLSPIEGELGLAS

ARMOUR PLATE GLASS — SICHERHEITSGLAS

CAST GLASS — GUSSGLAS

ROLLED GLASS — WALZGLAS

FIGURED ROLLED GLASS — ORNAMENTGLAS

REEDED GLASS — RILLENGLAS

CATHEDRAL GLASS — KATHEDRALGLAS

WIRED GLASS — DRAHTGLAS

BULLET-PROOF GLASS — PANZERGLAS, SCHUSS-SICHERES GLAS

CLADDING GLASS — VERKLEIDUNGSGLAS

GLASS BLOCKS — GLASBAUSTEINE

POLISHED SILVERED PLATE GLASS = MIRROR
SPIEGEL

CHROME PLATED DOME CAPPED SCREW
VERCHROMTE RUNDKOPFSCHRAUBE

GLAZING
VERGLASUNG

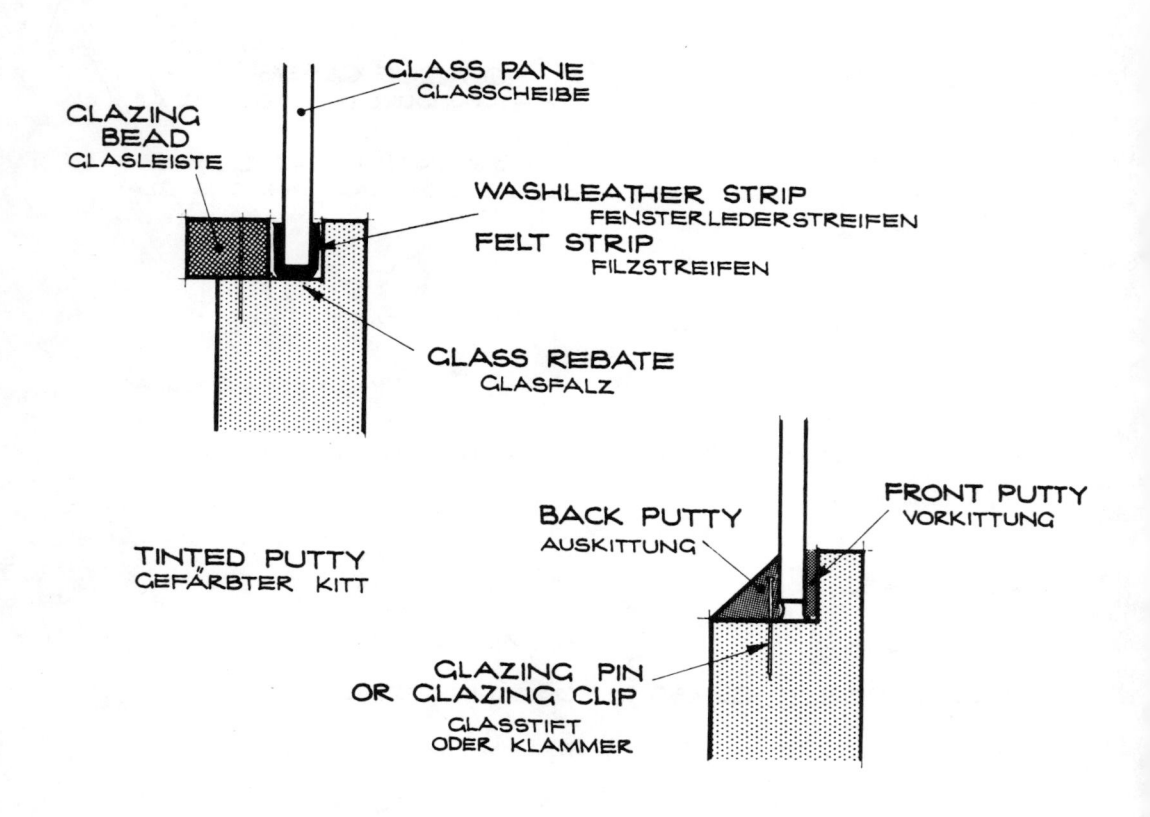

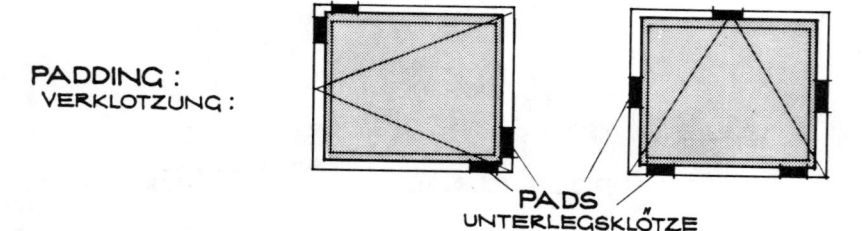

GLAZIER
GLASER

- GLASS PANE — GLASSCHEIBE
- GLAZING BEAD — GLASLEISTE
- WASHLEATHER STRIP — FENSTERLEDERSTREIFEN
- FELT STRIP — FILZSTREIFEN
- GLASS REBATE — GLASFALZ
- TINTED PUTTY — GEFÄRBTER KITT
- BACK PUTTY — AUSKITTUNG
- FRONT PUTTY — VORKITTUNG
- GLAZING PIN OR GLAZING CLIP — GLASSTIFT ODER KLAMMER
- PADDING: — VERKLOTZUNG:
- PADS — UNTERLEGSKLÖTZE

GLASS SPECIFICATION IN IMPERIAL UNITS : — ENGLISCHE GLASEINTEILUNG :

	IN WEIGHT (PER SQU. FOOT) NACH GEWICHT						IN INCHES NACH DICKEN		
IMPERIAL	12 oz	14 oz	16 oz	18 oz	24 oz	32 oz	3/16"	7/32"	1/4"
EQUIVALENT IN METRIC	1,2	1,6	1,8	2	3	4	5	5,5	6,5 mm

DICKEN IN MILLIMETERN

HINGES AND BOLTS
BÄNDER UND RIEGEL

IRONMONGER
BESCHLÄGESPEZIALIST

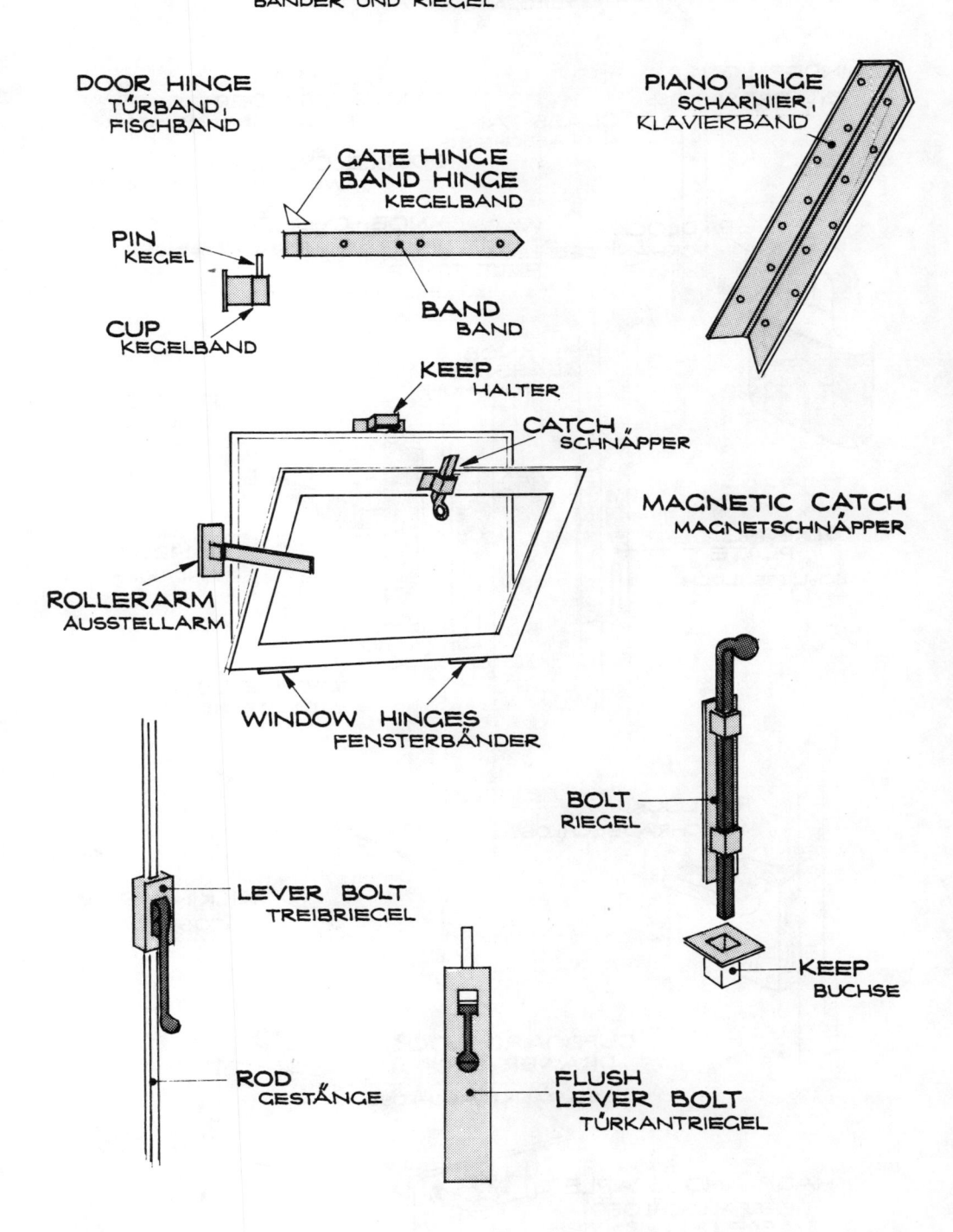

DOOR HINGE — TÜRBAND, FISCHBAND

PIANO HINGE — SCHARNIER, KLAVIERBAND

GATE HINGE / BAND HINGE — KEGELBAND

PIN — KEGEL

CUP — KEGELBAND

BAND — BAND

KEEP — HALTER

CATCH — SCHNÄPPER

MAGNETIC CATCH — MAGNETSCHNÄPPER

ROLLERARM — AUSSTELLARM

WINDOW HINGES — FENSTERBÄNDER

BOLT — RIEGEL

LEVER BOLT — TREIBRIEGEL

KEEP — BUCHSE

ROD — GESTÄNGE

FLUSH LEVER BOLT — TÜRKANTRIEGEL

159

LOCKS
SCHLÖSSER

IRONMONGER
BESCHLÄGESPEZIALIST

CYLINDER LOCK
ZYLINDERSCHLOSS

SINGLE CYLINDER
EINSEITIGER ZYLINDER

DOUBLE CYLINDER
ZWEISEITIGER ZYLINDER

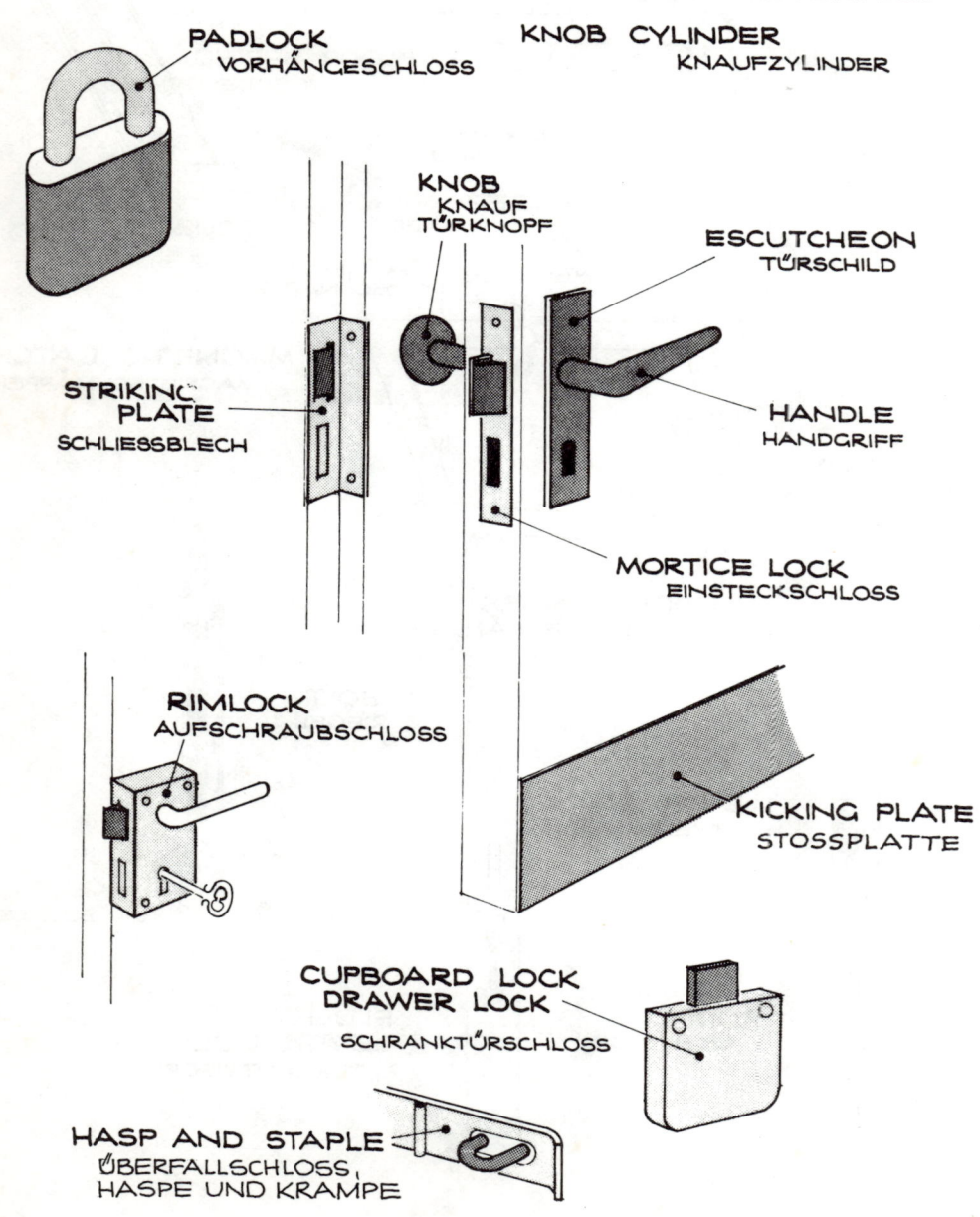

PADLOCK
VORHÄNGESCHLOSS

KNOB CYLINDER
KNAUFZYLINDER

KNOB
KNAUF
TÜRKNOPF

ESCUTCHEON
TÜRSCHILD

STRIKING PLATE
SCHLIESSBLECH

HANDLE
HANDGRIFF

MORTICE LOCK
EINSTECKSCHLOSS

RIMLOCK
AUFSCHRAUBSCHLOSS

KICKING PLATE
STOSSPLATTE

CUPBOARD LOCK
DRAWER LOCK
SCHRANKTÜRSCHLOSS

HASP AND STAPLE
ÜBERFALLSCHLOSS,
HASPE UND KRAMPE

VARIOUS
VERSCHIEDENES

IRONMONGER
BESCHLÄGESPEZIALIST

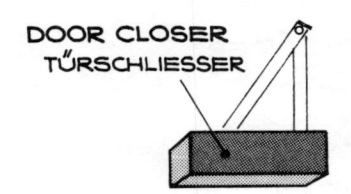

DOOR CLOSER
TÜRSCHLIESSER

FLOOR SPRING
BODENSCHLIESSER

SINGLE ACTION
FÜR DREHFLÜGEL

DOUBLE ACTION
FÜR PENDELFLÜGEL

STAY FASTENER
FESTSTELLER, AUSSTELLARM

DOOR STOP
TÜRSTOPP

COAT HOOK
KLEIDERHAKEN

IRONMONGERY MATERIALS:
  BESCHLÄGE - MATERIALIEN:

STEEL	BRASS	ALUMINIUM
STAHL	MESSING	

IRONMONGERY FINISHES:
  BESCHLÄGE - OBERFLÄCHEN:

SATIN NICKEL (SN)	SEIDENMATT VERNICKELT
CHROMIUM PLATED (CP)	VERCHROMT
POLISHED BRASS	POLIERTES MESSING
BRONZE PLATED	BRONZIERT

FLOOR FINISHES
BODENBELÄGE

FLOOR LAYER
BODENLEGER

HARD FINISHES:
HARTE BELÄGE:

CEMENT SCREED	ZEMENTESTRICH
EPOXY PAINT ON SCREED	EPOXYANSTRICH AUF ESTRICH
TERRAZZO	TERRAZZO
FLOOR TILES	BODENPLATTEN
PAVING SLABS	GEHWEGPLATTEN

TIMBER FINISHES:
HOLZBELÄGE:

TIMBER BOARDING	HOLZRIEMENBODEN
PARQUET BLOCKS	PARKETT
CHECKERED PATTERN	WÜRFELMUSTER
HERRING BONE PATTERN	FISCHGRÄTMUSTER

PLASTIC FINISHES:
PLASTIKBELÄGE:

VINYL TILES (PVC)	PVC-FLIESEN
VINYL-ASBESTOS TILES (VA)	PVC-ASBESTFLIESEN

SOFT FINISHES:
WEICHE BELÄGE:

CARPET	TEPPICH
WALL-TO-WALL CARPET FITTED CARPET	SPANNTEPPICH

WALL FINISHES
WANDBELÄGE

PAPERHANGER
TAPEZIERER

HARD FINISHES :
HARTE BELÄGE :   (BRICKLAYER)

    STONEWORK   NATURSTEIN
    FACING BRICKWORK   SICHTZIEGEL
    PLASTER   VERPUTZ
    PAINT ON PLASTER   ANSTRICH AUF VERPUTZ
    TERRAZZO   TERRAZZO
    WALL TILES   WANDFLIESEN

TIMBER FINISHES:
HOLZBELÄGE :   (CARPENTER)

    TIMBER CLADDING   HOLZVERKLEIDUNG
    TIMBER PANELLING   HOLZVERTÄFELUNG

PLASTIC FINISHES:
PLASTIKBELÄGE :   (PAPERHANGER)

    VINY WALL COVERING   PLASTIKWANDTAPETE

SOFT FINISHES :
WEICHE BELÄGE :   (PAPERHANGER)

    WALL PAPER   TAPETE
    HESSIAN   SACKLEINEN
    CANVAS   SEGELTUCH
    MATERIAL CLADDING   STOFFBESPANNUNG
    WALL CARPET,
    TAPESTRY, HANGINGS   WANDTEPPICH

PAINTER'S TOOLS
MALERWERKZEUG

PAINTER / MALER

- **SPATULA** — SPACHTEL
- **BRUSHES** — PINSEL, BÜRSTEN
- **BUCKET** — EIMER
- **LADDERS** — LEITERN
- **SANDPAPER** — SCHLEIFPAPIER
- **ADHESIVE** — KLEBER
- **PAINT ROLLER** — FARBWALZE
- **PLASTIC FILLER** — SPACHTELMASSE
- **PAINTS** — ANSTRICHFARBEN

TO PAINT STREICHEN LACKIEREN	TO OIL ÖLEN	BRUSH APPLICATION STREICHVERFAHREN	
TO STAIN BEIZEN	TO WAX WACHSEN	ROLLER APPLICATION AUFWALZVERFAHREN	
TO STOP TO FILL SPACHTELN	TO KNOT KITTEN	SPRAY APPLICATION SPRITZVERFAHREN	
	TO SANDPAPER SCHLEIFEN	DIP APPLICATION TAUCHVERFAHREN	

PAINT ON PLASTER
FARBE AUF VERPUTZ

PAINTER
MALER

PRIMER COAT:
GRUNDIERANSTRICH:

ALKALI - RESISTANT PRIMER
    LAUGENFESTES GRUNDIERMITTEL

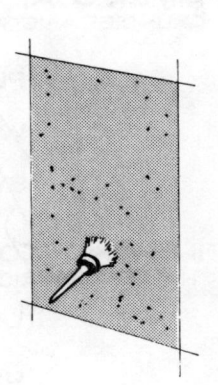

FINISHING COAT:
HAUPTANSTRICH:

PVA PAINT	PVA-FARBE
DISTEMPER	LEIMFARBE
LIME WASH	KALKFARBE
OIL PAINT	ÖLFARBE
POLYESTER PAINT	POLYESTERFARBE
ACRYLIC PAINT	ACRYLFARBE
EMULSION PAINT	EMULSIONSFARBE

PAINT ON TIMBER
FARBE AUF HOLZ

PAINTER
MALER

PRIMER COAT:
GRUNDIERANSTRICH:

    PLASTIC FILLER — SPACHTELMASSE

    WOOD PRIMER — HOLZGRUNDANSTRICH

    SYNTHETIC RESIN PRIMER — KUNSTHARZGRUNDANSTRICH

FINISHING COAT:
HAUPTANSTRICH:

    UNDERCOAT — VORANSTRICH

    ENAMEL PAINT — LACKFARBE

        SEMI-GLOSS ENAMEL — SEIDENGLANZLACK

        HIGH-GLOSS ENAMEL — EMAILLELACK

    LACQUER — KLARLACK

FINISH ON NATURAL TIMBER
BEHANDLUNG VON NATURHOLZ

PRIMER COAT:
GRUNDIERUNG:

    SANDING SEALER — VORLACK, SCHLEIFLACK

    STAIN — BEIZE

FINISHING COAT:
HAUPTANSTRICH:

    BEESWAX THINNED WITH TURPENTINE — MIT TERPENTIN VERFLÜSSIGTES BIENENWACHS

    VARNISH — FIRNIS

    POLYURETHANE — KUNSTSTOFF-FIRNIS

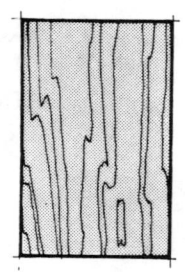

PAINT ON METAL
FARBE AUF METALL

PAINTER
MALER

PRIMER COAT:
GRUNDIERANSTRICH:

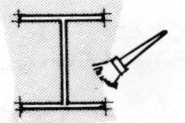

    ZINC CHROMATE PRIMER   ZINKCHROM - MENNIGE

    SYNTHETIC RESIN PRIMER
                                 KUNSTHARZMENNIGE

    RED LEAD PRIMER   BLEIMENNIGE

FINISHING COAT:
HAUPTANSTRICH:

    UNDERCOAT   VORANSTRICH

    ENAMEL PAINT   LACKFARBE

        SEMI - GLOSS ENAMEL   SEIDENGLANZLACK

        HIGH - GLOSS ENAMEL   EMAILLELACK

PAINT ON PVC
FARBE AUF PVC

PVC GUTTER

PVC - (BESCHICHTETE) REGENRINNE

PRIMER AND
FINISH:
VOR - UND HAUPTANSTRICH:

    ACRYLIC PAINT   ACRYLFARBE

Alphabetisches Verzeichnis der bautechnischen Begriffe
Alphabetical Index of Technical Terms in Construction

A

Abbindebehandlung	75	Anschlußplatte	152
Abbindwärme	75	Anschlußstück	149
Abbolzung	80	Ansicht	8, 9
Abbolzungsstütze	60	Anstrichfarben	164
Abdeckgitter	45	Antenne	145
Abdeckrost	57, 156	Apartment	4
Abdeckungen	90	Arbeiter, gelernter	34
Abhängungsstab	106	Arbeiter, ungelernter	34
Abkantung	41, 100, 108	Arbeitsraum	58
Ablage	122	Architektur	2
Ablagebretter	122	Asbestmühle	18
Ablagefach	122	Asbestzement	42
Ablauf	150	Asbestzementplatte	138, 142
Ablauföffnung	150	Asbestzementrohr	149
Ablaufpfropfen	150	Asbestzementschiefer	139
Ablaufrost	150	Attika	67
Ablegleiste	121	Attikabalken	62
Abluftkamin, dreizügiger	88	Aufbau	52
Abrechnung	5	Aufbereitungsanlage	146
Abreißbewehrung	76	Aufdoppelungsbrett	134
Abrichte	117	Auffahrrampe	68
Absatz	40	Auffüllung	58
Absatzstück	149	Auflager	28
Abschlagzahlung	7	Auflager, eingespanntes	28
Absetzbecken	155	Auflager, freies	28
Absolutkote	38	Auflagerdruck	24
Absperrhahn	150	Auflagerpressung	24
Absperrventil	146, 150	Auflast	26
Abstandhalter	81, 104	Aufmaß	11
Absteifung	80	Aufschraubschloß	160
Abstellflügel	128	Auftritt	66
Abteil	45	Aufwalzverfahren	164
Abwasser	154	Aufzug	50, 99
Abwicklung	108	Ausbau	7
Abzug	153	Ausbesserungsarbeiten	75
Abzugschacht	88	Ausblühung	43
Abzugskanal	156	Ausfachung	64
Abzweigleitung	146	Ausfachungsmauerwerk	64
Achsabstand	42	Ausfaulgrube	155
Achslinie	12	Ausführungszeichnung	6
Acrylfarbe	165, 167	Aushubarbeit	36
Änderung	11	Aushubmaterial	54
Afrormosia	115	Auskittung	158
Alkoven	15	Auskleidung	109
Aluminium	98	Auslaufstück	149
Aluminiumfolie	140	Ausleger	49
Amboß	99	Auslegerkran	50
Angebotseinholung	6	Ausmauerung	89, 153
Anker (Draht-)	89	Ausschalter	144
Ankerbolzen	106	Ausschalzeiten	80
Ankerloch	40	Ausschuß	114
Ankleideraum	15	Aussparungen	90
Anprall	27	Aussparungsplan	6
Anrichte, Buffet	122	Aussteifung	54, 105
Anschlag	88	Aussteifungsstab	105
Anschlußeisen	77	Aussteifungsstrebe	135
		Ausstellarm	159

Außengewinde	148
Außenputz	92
Außentür	132
Außenwand	52, 61
Automobilfabrik	17
Axt, Beil	118

B

Bad	153
Badewanne	153
Badsiphon	153
Bagger	49
Bahnen	142
Bahnstrecke	103
Balken	52
Balken, deckengleicher	62
Balken, eingespannter	62
Balken, freiaufliegender	62
Balkon	14
Balkonentwässerung	156
Band	159
Bandsäge	117
Bandstahl	101
Bankett	14, 55, 58
Bauabnahme	6
Bauarbeiten	47
Baubüro	48
Baugelände	14
Baugerüst	50
Baugrundstück	43
Bauhandwerker	35
Bauherr	34
Bauholz	37, 114
Bauingenieurwesen	19
Baukantine	48
Bauklammer	80
Baukosten	7
Baukunde	31
Baulabor	48
Baulager	48
Bauleitung	5
Bauleitung, örtliche	5, 6
Baustahlgewebe	76
Baustelle	14, 47
Baustelleneinrichtung	48
Baustellenräumung	48
Baustellenstraße	48
Baustellenüberwachung	47
Bauunterkünfte	48
Bauzeitplan	6
Bebauung	8
Bebauung, geplante	11
Bedienungsknopf	150
Befestigungslasche	104
Befestigungsschiene	106
Befestigungsschiene, einbetonierte	95
Befestigungswinkel	106
Beilagscheibe	104
Beistelltisch	121
Beize	166
Belastung, tatsächliche	25
Belastung, zulässige	25
Belastungsfälle	26
Beleuchtungskörper	144
Bemessung	21
Benzinabscheider	155
Bepflanzungsreste	53
Berechnung, statische	6, 21
Beschlagarbeit	36
Beschlägefachmann	137
Beschlägespezialist	35
Beton, unbewehrter	71
Betonarbeit	36, 70
Betonblock	85
Betondeckel	155
Betonfertigteil	42
Betonbauer	35
Betonieranlage	49
Betonmischer	49
Betonrohr	149
Betonrüttler	50
Betonspannung	74
Betonstahlbieger	35
Betonstahlverleger	35
Betonüberdeckung	76
Bettgestell	121
Bewässerungsanlage	146
Bewehrung	70
Bewehrung, obere	78
Bewehrung, untere	78
Bewehrungsarbeit	36
Bewehrungskorb	76
Bewehrungsplan	6
Bewehrungsstab	76
Bewehrungsstahl	37
Bewehrungsstange	76
Bezug	11
Bezugszeichnung	11
Biberschwanzziegel	138
Bidet	152
Biegeknickung	29
Biegeliste	78
Biegemoment	24
Biegsamkeit	23
Biegezugfestigkeit	24
Biegung	29
Bindedraht	78
Binder	67, 105
Binder, falsche	91
Binderverband	87
Bitumenpappe	141
Blasebalg	99
Blatt, Verblattung	116
Blaupause	12
Blech	101
Blei	98
Bleimennige	167
Blende, Vorhangleiste	130
Blendmauer	61, 89
Blocklast	26
Boden	123

Boden, gewachsener	53
Bodenablauf	156
Bodendiele	80
Bodendurchbruch	40
Bodenebene	52
Bodenkanal	40
Bodenkanal, offener	57
Bodenleger	35
Bodenlegerarbeit	36
Bodenplatte	13, 55, 58, 94, 95
Bodenplattenbeläge	93, 162
Bodenpressung	24
Bodenrinne	156
Bodenrinne, halbkreisförmige	57
Bodenschacht	154
Bodenschließer	161
Bodenschlitz	40
Bodentür	132
Bogen	89, 149
Bogensäge	118
Bohle	114
Bohrer	99
Bohrmaschine	99, 118
Bolzen	80, 104
Bolzen, hochfester	104
Bolzenkopf	104
"Bosch"-Hammer	50
Brandmauer	61
Brauchwasser	154
Brausekopf	153
Brecheisen	84
Brett	114
Bronze	98
Brüstung	14, 52
Brüstungsmauer	61, 130
Buchse	159
Bude	4
Bücherbrett	122
Bücherregal	122
Bügel	77
Bürogeschoß	16
Bürstauftrag	141

D

Dachablauf	156
Dachablaufanschlußblech	140
Dachbinder	105
Dachdeckerarbeit	36
Dachentwässerung	156
Dachfenster, Gaupen	127
Dachflächenfenster	127
Dachgeschoß	14
Dachhaut	124, 140
Dachkleber	137
Dachneigung	14, 65
Dachpappe	37, 141
Dachpfanne	138
Dachplatte	95
Dachrinne	126
Dachschiefer	139
Dachschindel	139
Dachstuhl	67
Dachüberstand	67
Dachvorsprung	67
Dachziegel	37, 138
Dämmerungsschalter	144
Damm	56
Dampfsperre	140
Decke, abgehängte	106
Decke, kreuzweise gerippte	63
Deckel	123
Deckenbalken	62
Deckenbretter	126
Deckendurchbruch	40
Deckenlattung	126
Deckenplatte	13, 52, 63, 140
Dehnfugenabdichtung	140
Dehnung, Ausdehnung	29
Dehnungsfuge	39
Detailentwurf	6
Detailplan	6
Diagonalstäbe	105
Dichte	23
Dichtungsarbeit	36
Dichtungsmasse	37
Dichtungsring	148
Dichtungsstrick	130, 142, 148
Diele	15
Dimmer	144
Doppelabzweig	149
Doppelbiegung	29
Doppelfalz	108
Doppelstrangsystem	154
Doppel-T-Profil	101
Doppel-T-Stahl	101
Doppel-T-Stück	149
Doppelzimmer	15
Dränung	51
Draht	47, 102
(Draht)-Anker	89
Drahtglas	157
Drehbank	99, 117, 147
Drehflügel	128
Drehflügeltür	132
Drehkippflügel	128
Drehtür	132
Dreieck	22
Dreieckslast	26
Dreiecksleiste	81
Dreiphasenstrom	143
Dreiphasen-Wechselstrom	42
Druck	27
Druckfestigkeit	74
Druckbewehrung	76
Druckerei	17
Druckstab	105
Dübel	95, 116
Düker	156
Durchbiegung	29, 98
Durchbruch	40

Durchführbarkeitsstudie	6
Durchgang	15, 16
Durchlaufbalken	62
Durchlauferhitzer	146, 153
Dusche	15

E

Ecke, äußere	88
Ecke, innere	88
Eckleiste	13, 120, 126
Eckschutzleiste	120
Ecksichtstein	85
Eibe	113
Eiche	113
Eigengewicht	25
Eigenheim	4
Eimer	84
Einbauanweisung	45
Einbaurahmen	129, 130
Einbauschrank	122
Einfachabzweig	149
Einfahrtstor	135
Einfriedungsmauer	61
Eingang	15
Einkornbeton	71
Einlauf	155
Einlaufschacht	154
Einlegleiste	120
Einschalung	80
Einschlagpfosten	134
Einspannlänge	28
Einspannmoment	24
Einsteckschloß	160
Einstrangsystem	154
Einsturz	43
Einfeldbalken	62
Einzelfenster	127
Einzelfundament	55
Einzellast	26
Einzelstufen	66
Einzelzimmer	15
Einzimmerwohnung	4
Eisen	98
Eisensäge	147
Elastizitätsmodul	23
Elektriker	35
Elektroinstallation	36
Elektromotor	50, 145
Elektrosockelleiste	145
Emaillelack	166, 167
Emulsionsfarbe	165
Emulsionshaut	141
Endauslauf	110
Endvorkopf	110
Endzahlung	7
Energie	18
Entlüftungsrohr	155
Entlüftungsstrang	154
Entwässerung	54
Entwässerungsleitung	156
Entwurf	5
Entwurf, konstruktiver	21
Epoxyanstrich	162
Epoxyhaut	141
Erdarbeiten	54
Erddamm	56
Erdgeschoß	14
Erdkabel	143
Erdung	143
Erholungszentrum	4
Erle	113
Errichtung	43
Erstellungskosten	7
Erweiterungsraum, -gebiet	17
Esche	113
Espe	113
Esse	99
Eßzimmer	15

F

Fabrik	17
Fachwerk	64
Fachwerkträger	105
Fallbär	49
Fallrohr, rechteckiges	110
Fallrohr, rundes	110
Fallrohrauslauf	110
Faltfenster	128
Falttür	132
Falz	39, 100
Falzhobel	117
Falzverbindung	116
Falzziegel	138
Farbe	37
Farbglas	157
Farbwalze	164
Faserplatte	115
Fassade, vorgehängte	95
Fassadenbalken	62
Faulgrube	155
Feder	39
Feile	99
Fels, loser	53
Fels, massiger	53
Fels, verwitterter	53
Fenster	112
Fensterband	127, 159
Fensterbank	90, 130
Fensterbankisolierung	90
Fensterblech, äußeres	109
Fenstergitter	131
Fensterladen	81, 131
Fensterlederstreifen	158
Ferienhaus	4
Fertigbeton	71
Fertigkote	38
Festigkeit	23, 24
Feststeller	161
Festverglasung	128
Fettabscheider	155
Feuchtigkeitsisolierhaut	58
Feuchtigkeitsisolierschicht	90

Feuchtigkeitsisolierung, vertikale	58
Feuchtigkeitssperre	52, 58
Feuerfluchtweg	16
Feuerstelle	88
Feuerwand	61
Fichte	113
Fichtenholz, geschnittenes	114
Filzstreifen	158
Finanzierungsplan	6
Findlingsgestein	53
Firnis	166
First	67, 125
Firstpfette	124
Fischband	159
Fischgrätmuster	162
Flachasbestplatte	138
Flachdach	65
Flachdachentwässerung	156
Flachnaht	104
Flachstahl	101
Fläche	23
Fläche, abgewinkelte	108
Fläche, überdachte	15
Fläche, überbaubare	7
Fläche, überbaute	7
Fläche, überdachte	7
Flächenlast	26
Flansch	101
Flanschblech	101
Flaschenrüttler	75
Flaschensiphon	150, 151
Flaschenzug	50, 99
Flecken	98
Fliese	37
Fliese, keramische	94
Fliesenanschlußblech	151
Fliesenarbeiten	83
Fliesenleger	35
Fliesenlegerarbeit	36
Flocken	142
Flügelrahmen	129, 130
Flur	15
Fluß-Stahl	37, 42
Folie, wärmeisolierende	142
Fräse	118
Freisitz	15
Frischwassergerät	151
Frisiertisch	122
Froschperspektive	9
Fuchsschwanzsäge	117
Fuge	87
Fuge, ausgekratzte	87
Fuge, geschlossene	119
Fuge, offene	39, 119
Fuge, versetzte	87
Fugenband	39
Fugenklotz	94
Fugenlehre	91
Fundamentabstufung	55
Fundamentbeton	71
Fundamente	14
Fundamentisolierung	90
Fundamentplatte	55
Fundamentsockel	68
Fundamentsohle	55
Furnier	115
Furnierpresse	117
Fußbodenoberkante	38
Fußpfette	124
Fußplatte	106
Fußschiene	123
Futter	104, 120, 133
Futterleiste	120

G

Gabelstapler	17
Gang	15
Ganzglastür	132
Garage	15
Garagentor	135
Garderobe	16, 122
Gartenarchitektur	3
Gartentor	135
Gasbeton	95
Gasbetonblock	85
Gebäude	14
Gebäude, mehrgeschossiges	14
Gebrauchsklasse	114
Gefälle, Böschung	41, 53
Gefällebeton	41, 71, 140
Gegengewicht	27, 50
Gehwegplatten	162
Gelände, flaches	53
Geländer	14, 66, 107
Geländerpfosten	107
Gelenk	64
Gemeinschaftswand	61
Generalplan	6
Geometrie	22
Geometrie, darstellende	21
Geräteraum	15
Gerippe	64
Geröll	53
Geruchverschluß	155
Gesamtgrundriß	8
Gesamtlänge	28, 42
Geschäftshaus	4
Geschäftsräume	4
Geschoß	14
Geschoß, oberes	14
Geschoß, unteres	14
Geselle	34
Gesimsarbeit	92
Gestänge	159
Gestein	73
Gewicht	23
Gewinde	148
Gewindeschneider	147
Gewölbe	89
Giebel	67
Giebelabdeckung	109

Gipsarbeit	92	Halle	17
Gitterrost	102	Hallenbad	4
Gitterabfüllung	107	Halteblech	106
Glättscheibe	84	Halter	159
Glas, lichtdurchlässiges	157	Handbohrer	117
Glas, undurchsichtiges	157	Handelsklasse	114
Glasablage	151	Handgriff	160
Glasbaustein	157	Handlauf	66, 107
Glasdachziegel	138	Handlaufhalterung	107
Glaseinteilung, englische	158	Handsäge	117
Glaser	35, 137	Handtuchhalter	94
Glasarbeit	36	Handtuchstange	151
Glasfalz	158	Handverdichtung	74
Glasfaserdachziegel	138	Handwaschbecken	151
Glasfaserlichtkuppel	140	Hartfaserplatte	115
Glasfaserplatte	142	Hartholz, Laubholz	113
Glasleiste	158	Hartholzdübel	116
Glasscheibe	158	Haspe	160
Glasschrank, Vitrine	122	Hauptanstrich	165
Glasstift	158	Hauptbebauungsplan	6
Glaswollbahn	142	Hauptgebäude	17
Gleichgewicht	27	Hauptstiel	124
Gleichlast	26	Hauptunterbrecher	143
Gleichstrom	42	Hauptunternehmer	34
Gleis	103	Hauptverteilerkasten	143
Gleisanschluß	103	Hauptwasserleitung	146
Gleitschalung	81	Hausarbeitsraum	15
Gleitschutz	66	Hebeanlage	146
Gleitsicherheit	29	Hebepfropfen	150
Glühbirne	144	Hebetür	132
Graben	54	Hecke	15
Grasdach	139	Heißauftrag	141
Grat	67, 125	Heißbitumen	141
Greifer	49	Heizgerät	145
Grenzmauer	61	Herd	145
Größen, statische	24	Hilfsarbeiter	34
Größtkorn	73	Hilfsöffnungen	91
Grubensand	73	Hinterhof	15
Gründung	55	Hobelbank	117
Grund, gewachsener	53	Hobelmaschine	117
Grundieranstrich	165, 166	Hobelspäne	114
Grundiermittel, laugenfestes	165	Hochbau	20
Grundierung	166	Hochbauarchitektur	3
Grundlagenzeichnung	11	Hochbaurahmen	64
Grundriß	8	Hochhaus	4
Grundstück	43	Hochkantschicht	87
Grundstücksgrenze	15, 18	Hochlochziegel	85
Grundwasser	52, 54	Hochlöffelbagger	49
Grundwasserspiegel	54	Hochspannung	143
Gußeisen	42	Hochspülkasten	152
Gußglas	157	Höhe, lichte	28
Gußrohr	149	Hohlblock	85
Gutachten	6	Hohlblockstein	37
		Hohlfuge	87
H		Hohlkehle	41, 92
Haarnadel	77	Hohlnaht	104
Hängefach	122	Hohlprofilstütze, rechteckige	106
Hängewerk	124	Hohlraum	89
Haftgrund	91	Hohlträger	105
Haftung	27	Holm	107
Haken	50	Holz, Brennholz	114

Holz, baumkantiges	114
Holz, fehlkantiges	114
Holz, harziges	113
Holz, leimgebundenes	115
Holz, scharfkantiges	114
Holzarten	112
Holzfaserplatte	115
Holzfußboden	13
Holzgrundanstrich	166
Holzhobel	117
Holzpflaster	93
Holzriemenboden	162
Holzschindel	139
Holzscheibe	84
Holzschraube	116
Holzsockelleiste	13
Holzstütze	60
Holzverkleidung	163
Holzvertäfelung	163
Holzwolle	142
Horizontalschubkomponente	27
Hütte	4
Humus, Mutterboden	53

I

Industrieanlage	18
Industrieansiedlungszone	7
Industriebau	17
Industriegebiet	7
Innenarchitektur	3
Innengewinde	148
Innenputz	92
Innenrüttler	75
Innentür	132
Innenwand	61
Installateur	35, 137
Isolator	143
Isolierarbeit	36
Isolierer	137
Isolierhohlblock	142
Isoliermaterial	37
Isolierplatte, mehrschalige	95
Isolierwand	89

J

Jalousie	131
Junggesellenwohnung	4

K

Kabel	143
Kabelrohr	143
Kabine	45
Kalkfarbe	165
Kalkmörtel	86
Kalkputz	92
Kalkputz, zweilagiger	92
Kalkzementmörtel	86
Kaltauftrag	141
Kaltwasserversorgung	146
Kamin	13, 88
Kamin, offener	88
Kamineinfassung	109
Kanal	40, 52, 57
Kante, abgerundete	119
Kante, gebrochene	119
Kantenschutzeisen	92
Kantholz	114, 120
Kartonpapier	12
Kastenrinnen	110
Kastenträger	105
Kathedralglas	157
Kegel	22
Kegelband	159
Kehlbalken	124
Kehlblech	109
Kehle	67, 125
Kehlnaht	104, 106
Kehlsparren	125
Keil	80
Kelle	84
Kellerbar	15
Kellergeschoß	14
Kellergeschoß, unteres	14
Kerbsäge	117, 118
Kiefer, Föhre	113
Kies	37
Kies, gebrochener	73
Kies, natürlicher	73
Kieselsteine	73
Kiesfüllung	58
Kiesnester	75
Kippflügel	128
Kippsicherheit	29
Kitt	37
Klappflügel	128
Klarglas	157
Klarlack	166
Klavierband	159
Kleber	164
Kleberdübel	95
Kleiderhaken	161
Kleiderstange	122
Kleinstkorn	73
Klemme	143
Klinker	85
Knauf, Türknopf	160
Knaufzylinder	160
Knicklänge	28
Knickung	29
Knotenblech	104, 106
Knotenpunkt	104
Köcherfundament	95
Kompressor	50
Konsistenz	23, 74
Konsole	63
Kontrollschacht	154, 156
Kopfbug	105
Kopfplatte	106
Korkplatte, expandierte	142
Korngröße	73
Korntrennung	45, 73
Korrektur	9

Korrosion	98
Kostenschätzung	5
Kote	38
Kraftangriff	27
Kräfte	27
Kragarm	63
Kragarmlänge	28
Krageisen	63
Kragplatte	63
Krampe	160
Kranbahn	103
Kranwagen	50
Kreis	22
Kreiselpumpe	49
Kreissäge	118
Krempziegel	138
Kreuzbogen	81
Kreuzgewölbe	89
Kreuzschalter	144
Kreuzung	103
Kristallspiegelglas	157
Krümmer	120
Kunstharzgrundanstrich	166
Kunstharzmennige	167
Kunstschmiedearbeiten	107
Kunststein, Terrazzo	93
Kunststeinarbeit	36
Kunststoffolie	140
Kupfer	98
Kupferblech	108, 138
Kupferrohr	149
Kupferschindel	139
Kuppel	65

L

Lackfarbe	166, 167
Laden	81
Laderampe	68
Längseisen	77
Längsschnitt	8
Längsdruck	27
Längskraft	27
Längssichtstein	85
Lärche	113
Läuferschicht	87
Läuferverband	87, 91
Lageplan	8
Lagerbehälter	48
Lagerfuge	87
Lagerhölzer	80
Lampe	144
Landesgrenze	15
Landschaftsarchitektur	3
Langlochziegel	85
Langrundloch	106
Lasche	80, 116
Lastannahme	25
Lastarten	25
Latte	114, 120
Laube	15
Laubholz	113

Laubsieb	156
Laufkatze	50
Laufkatzengleis	103
Laufplatte	66
Lautsprecheranlage	145
Legende	11
Lehm	53
Lehne	121
Leibung	130, 133
Leichtbauplatte	85
Leimbinder	115
Leimfarbe	165
Leiste	120
Leistungsverzeichnis	6
Leiter	164
Leiter, isolierter	143
Leiter, nichtisolierter	143
Leitungskanal, begehbarer	57
Leitungsschlitz	40, 57
Leuchtdecke	144
Leuchte	153
Leuchtröhre	144
Leuchtstoffröhre	144
Lichtraumprofil	103
Linde	113
Linienlast	26
Lochziegel	85
Löffel	49
Lötkolben	100
Lötlampe	100, 147
Löhtnaht	108
Lötzange	100
Lötzinn	100
Lot, Senkblei	84
Lüftungslamellen	131
Luftschutzraum	15

M

Magerbeton	71
Magnetschnäpper	159
Mahagoni	115
Maler	35, 137
Malerarbeit	36
Manschette	148
Mantelfläche	108
Mantelreibung	56
Markise, Sonnenblende	131
Maschendraht	102
Maschinenfundament	55
Massenberechner	5
Massenbeton	71, 75
Massenermittlung	5
Maßtoleranzen	81
Mast	60
Maßaufnahme	11
Materialeigenschaften	23
Mauerabdeckung	90, 109
Mauerecke	88
Mauerfundament	55
Mauersockel	68
Mauerziegel, gewöhnlicher	85

Maurer	35
Maurerhammer	84
Medikamentenschränkchen	90
Mehrfamilienhaus	4
Meißel	84
Messing	98
Metallrahmen	133
Metallzarge	133
Mineralwolle	142
Mischbatterie	150
Mittelauslauf	110
Mittelpfette	124
Möbel	112
Möbelschreiner	121
Mörtelbett	91
Mörtelbrücken	91
Mörtelreste	91
Moment	24
Montage	43
Montagebewehrung	77
Montagebock	77
Montagedeckel	130
Montageeisen	77
Muffe	148
Mutter	104
Mutterpause	12

N

Nachttisch	121
Nadelholz	113
Nagel	116
Nageleisen	118
Nase, Überstand	66
Naturschiefer	139
Naturstein	85, 163
Naturzement	72
Nebengebäude	17
Nebengleis	103
Nebenverteilerkasten	143
Netzbewehrung	76
Nirosta	151
Nivelliergerät	47
Norden	12
Normalspur	103
Normenzement	72
Nußbaum	113
Nut	39
Nutzfläche, gewerbliche	7
Nutzlast	25

O

Oberbau	14, 52
Oberflächenentwässerung	156
Oberflächenrüttler	75
Oberflächenwasser	54
Obergeschoß	14
Obergurt	105, 124
Oberkante	38
Oberleitungs-Verteilernetz	143
Oberlicht	127
Oberlichtanschlußblech	140

Ölfarbe	165
O. M.	10
Ornamentglas	157
Ortbeton	71
Ortgang	67
Ortgangbrett	67
Ortterrazzo	93

P

Panzerglas	157
Papierfabrik	17
Papierhalter, eingelassener	94
Papierpause	12
Pappe, imprägnierte	141
Pendeltür	132
Perspektive	9
Pfahl	47, 56
Pfahlfuß	56
Pfahlkopf	56
Pfeiler	60, 88
Pflaster	83, 93
Pflaster, säurefestes	93
Pfosten	60
Pickel	84
Pilzdach	65
Pilzdecke	63
Pinie	113
Pinsel	164
Pissoirbecken	152
Pissoirrinne	152
Pissoirschüssel	152
Plan	8
Planierraupe	49
Planung	5
Plastikhaut	141
Plastikpause	12
Plastikrohr	37
Plastikschlauch	150
Plastikwandtapete	163
Platten	142
Plattenbalken	62
Podest	66
Podestplatte	66
Polster	121
Polyäthylenfolie	140
Polyesterfarbe	165
Portlandzement	72
Porzellanfliese, glasierte	94
Prallwand	155
Preßlufthammer	50
Privathaus	15
Probewürfel	74
Produktionshallen	17
Profillehre	45, 84
Profilstütze, zusammengesetzte	60
Projekt, schlüsselfertiges	7
Projektüberwachung	5
Prospekt	45
Pultdach	65
Punktlast	26
Punktstrahler	144

Putz	92
Putzdecke	13
Putzflügel	128
Putzkelle	92
Putzleiste	133
Putzstreifen, Schürze	109
PVA-Farbe	165
PVC-Asbestfliese	162
PVC-Bahn	140
PVC-Deckprofil	107
PVC-Fliese	162
PVC-Rohr	149
Pyramide	22

Q

Quadrat	22
Quadratrohr	102, 148
Querband, Windrispe	125
Querholz	134
Querkraft	24
Querriegel	62
Querschnitt	8, 76
Querschnittsfläche	23
Querstrebe	105

R

Rabitz	102
Rabitz-Arbeit	92
Radlast	25
Rahmen	134, 135
Rahmenoberteil	134
Rahmenpfosten	134
Rahmentragwerk	64
Rahmenunterteil	134
Rahmenwerk	64
Ramme	49
Rand	45
Randbalken	62
Randdifferenz	45
Randstreifen	109
Rangiergleis	103
Rasen	15, 18
Rasiersteckdose	153
Raster	12
Rasterlinie	12
Rauchfang	99
Rauldübel	95
Rauminhalt	7
Raupe	49
Raupenkette	49
Rechenschieber	21
Rechteckrohr	102, 148
Reduzierstück	149
Reetdach	139
Regal	122
Regenrinne, halbrunde	110
Regenrinne, PVC-beschichtet	167
Reibung	27
Reihenhaussiedlung	4
Reinigungsöffnung	155
Reinigungsstück	149

Rethdach	139
Reparaturlast	25
Richtfest	44
Riegel	64, 159
Riffelblech	102
Rille	57
Rillenglas	157
Ringanker	62
Rinne	54
Rinnenboden	110
Rinnhaken	110
Rippendecke	63
Riß	43
Rödeldraht	81
Röhre	144
Rohbau	7
Rohkote	38
Rohr	102
Rohrbiegemaschine	147
Rohreinfassung	109
Rohrgerüst	60
Rohrhalter	110
Rohrkern	148
Rohrleger	137
Rohrschraubstock	147
Rohrstütze	60
Rolladen	81, 131
Rolladenkasten	130
Rolladenschiene	130
Rolladentür	132
Rollbahnen	20
Rollkies	73
Rosette	150
Rost	98, 156
Rücksprung	40
Rückstauverschluß	156
Rückwand	123
Rüttelbohle	75
Rüttler	74
Rundholz	115
Rundkopfschraube, verchromte	157

S

Sackkalk	86
Sackleinen	163
Sägemehl	114
Sägespäne	114
Säule	60
Salzwasser	146
Sammelgrube	155
Sandfang	155
Sanitärabfallverbrenner	152
Sanitärinstallation	36
Satteldach	65
Sauberkeitsschicht	58
Segeltuch	163
Seidenglanzlack	166, 167
Seifenschale	94
Seifenschüssel	153
Seitenflügel	127
Seitenwand	123

Senkrechte	88	Sprüher	152
Sheddach	65	Sprühkopf	150
Sheddach-Konstruktion	17	Spülbecken	151
Sicherheitsglas	157	Spüle	151
Sicherheitsleuchte	144	Spülsiphon	151
Sicherung	143	Spülleitung	152
Sichtbeton	71	Spundwandgründung	56
Sichtfuge	87	Spurweite	103
Sichtmauerstein	85	Subunternehmer	34
Sichtziegel	163	Süßwasser	146
Sickeranlage	154	Summer	145
Sickerdole	154	Systemuntersuchung	21
Sickergrube	154		
Sitz	121	**SCH**	
Sitzbank	121	Schalarbeit	36, 80
Skelett	64	Schalblech	81
Skizze	5	Schale, äußere	89
Sockel	52, 123	Schale, innere	89
Sockelbalken	62	Schalendachkonstruktion	65
Sockelhöhe	91	Schalplan	6
Sockelleiste	13, 68, 93, 120	Schaltafel	81
		Schalter	144
Sockelverkröpfung	110	Schalter, Theke	123
Sofa	121	Schalung	70, 80
Sog	27	Schalung, stehende	81
Sohlenkote	155	Schalung, verlorene	81
Sommerhaus	4	Schalwerkzeug	81
Spachtel	84, 164	Scharnier	159
Spachtelmasse	164, 166	Schaufel	49, 84
Spannbeton	71	Schaufenster	129
Spanndraht	81	Scheiteldruck	27
Spanne	45	Schemel, Hocker	121
Spannschloß	81	Schenkel	101
Spannteppich	162	Scherlast (abscheren)	24
Spannung	24	Schicht, Lage	45
Spannungsermittlung	24	Schicht, stehende	87
Spannungsverteilung	24	Schiebefenster	128
Spannweite	28	Schiebeflügelrahmen	129
Spanplatte	115	Schiebetür	132
Sparren	124	Schieferdecker	137
Speicher	15	Schieferplatte	94
Speier	110	Schimmel	43
Speierauskleidung	110	Schlackenbeton	71
Speisekammer	15	Schlackenwolle	142
Speisezimmer	15	Schlackenzement	72
Spengler	35	Schlag	27
Spenglerarbeit	36, 97	Schlauchanschluß	150
Sperrholz	45	Schleiflack	166
Sperrholztürblatt	134	Schleifmaschine	118
Spezialrohrkupplung	148	Schleuse (Gang)	17
Spiegel	151	Schließblech	160
Spiegelschrank	94, 153	Schlosser	35
Spiralbewehrung	77	Schlosserarbeit	36, 97
Spitzende	148	Schlüsselschraube	116
Splint	104	Schmalspur	103
Sportzentrum	4	Schmiede	99
Spreizdorn	104	Schmiedeeisen	107
Sprengwerk	124	Schmutzwasser	154
Spritzputz	92	Schmutzwasserstrang	154
Spritzverfahren	164	Schnäpper	159
Sprühauftrag	141	Schnecke	107

Schneelast	25
Schneidemaschine	99
Schnitt	42
Schnittholz	112
Schnittkräfte	21
Schnurgerüst	47
Schrank	122
Schranktürschloß	160
Schraubstock	147
Schreiner	35
Schreinerarbeit	36, 112
Schub	27
Schubbewehrung	76
Schubfestigkeit	24
Schubkarren	84
Schublade	123
Schubladenfront	123
Schubspannung	24
Schubwiderstand	24
Schutzdach	14, 65
Schutzrohr	57
Schwanenhalsrohr	110
Schwedenschnitt	92
Schweißbrenner	99, 147
Schweißnaht	104
Schwellast	26
Schwelle	134
Schwenkarm	153
Schwenk-Auslegerkran	50
Schwergewichtsmauer	56
Schwerpunkt	23
Schwimmbagger	49
Schwindversuch	74
Schwingflügel	128
Schwingtor, Kipptor	135
Schwund	29

ST

Stabplatte	115
Stahl	98
Stahl, geschmiedeter	97
Stahl, verzinkter	42
Stahlbalken	106
Stahlbeton	42, 71
Stahlblech	108
Stahlblech, verzinktes	108, 138
Stahlhobel	117
Stahlkonstruktionen	97
Stahlliste	78
Stahlrippendecke	63
Stahlrohr	37, 149
Stahlrohr, verzinktes	149
Stahlstütze	106
Stahlstütze, leichte	60
Stahlstütze, schwere	60
Standhahn	150
Standrohr	150
Standsicherheit	29
Standsockel	68
Stapel	17
Starkstrom	143

Statik	21
Staudamm	56
Staumauer	56
Steckdosengehäuse	145
Steckeisen	77
Steg, Stegblech	101
Stehbügel	77
Steigeisen	102
Steigleiter	102
Steigung	66
Steine, künstliche	85
Steinpflaster	93
Steinschindel	139
Steinwolle	142
Stemmeisen	118
Stiel	64, 124
Stirnbrett	67
Stockwerk	14
Stoffbespannung	163
Stoß	104
Stoßblech	104
Stoßfuge	39, 87
Stoßplatte	160
Strebe	64, 80, 105, 124
Strebepfeiler	60
Streckenlast	26
Strecker	91
Streckmetall	102
Strecksäge	117
Streichverfahren	164
Streifenfundament	14, 52, 55
Strohdach	139
Stromkreis	143
Stromleiter	143
Stromspannung	143
Stuckarbeit	92
Stütze	52
Stützmauer	56
Studierzimmer	15
Sturz	62
Sturz, gemauerter	89
Sturzbalken	62
Sturzisolierung	90
Styropor-Platte	142

T

Tabelleneisen	78
Tafelglas	157
Tafel	142
Tanne	113
Tapete	13, 163
Tapeziererarbeit	36
Tapezierer	35, 137
Tauchverfahren	164
Teerisolierhaut	141
Teich	18
Teilgrundriß	8
Telefonzelle	45
Teppich	162
Terrasse	15

Terrassenentwässerung	156
Terrassenhaus	4
Terrazzo	162
Terrazzofliese	93, 94
Terrazzoplatte	93
Theodolit	47
Thermostat	145
Tiefbau	20
Tieflöffelbagger	49
Tiefspülkasten	152
Tiroler Rauhputz	92
Toilettenpapierhalter	152
Toilettenschrank	94
Toilettenschüssel	152
Toleranz	45
Ton	73
Tonrohr, Steinzeugrohr	149
Torfmull	142
Torkretier-Maschine	49
T-Profil	101
Träger	52
T-Stahl	101
T-Träger	67
Trägheitsmoment	23
Trägheitsradius	23
Tragkonstruktionen	20
Tragrahmen	64
Transparentpapier	12
Transportbeton	71
Trapez	22
Traufblech	109
Traufbrett	67, 126
Traufe	67
Traufel	84
Traufplatte	125
Trauflatte	126
Traverse, obere	107
Traverse, untere	107
Treibriegel	159
Trennwände	123
Treppe, Treppenhaus	52, 66
Treppenstufen	66
Treppenwange	120
Trittplatte, gerillte	94
Trommel	49
Türband	159
Türblatt	133
Türkantriegel	159
Türschild	160
Türschließer	161
Turmdrehkran	50

U

Überarbeitung	9, 12
Übderdeckungsstoß	78
Überfalle	160
Überlaufausschnitt	110
Überlaufrohr	146
Übersichtsplan	6, 8
Überwachung	5
Überzug	62

Ulme, Rüster	113
Umfassungsbalken	62
Umfassungswand	58, 61
Umleghebel	150
Umleimer	115, 134
Ungenauigkeiten	81
Unterbau	14, 52
Unterbeton(schicht)	58
Untergurt	105, 124
Unterkante	38
Unterlage	14
Unterlegfolie	139
Unterlegklotz	79
Unterputz-Mischbatterie	153
Untersicht	63
Untersichtkote	38
Unterzug	62
U-Profil	101
U-Stahl	101

V

Ventilator	145
Verankerungseisen	77
Verbindungseisen	77
Verbindungsmittel	97, 112
Verblendstein	85
Verblendungen	91
Verbohrung	54
Verbrennungsmotor	50
Verdichtung	74
Verformung	29, 98
Vergabe (von Aufträgen)	5
Verglasung	130
Verkehrslasten	25
Verkeilung	95
Verkleidung	133
Verkleidungsglas	157
Verklotzung	158
Vermittlungsapparat	145
Verputz	13, 83, 92, 163
Verputzarbeit	36
Verputzer	35
Verputzmaschine	50
Verrödelung	81
Verschalung	80
Versetzarbeiten	90
Verstemmeisen	147
Verstrebung	54, 80
Vertäfelung	13
Verteilereisen	77
Verteilerkasten	90
Vertikalschubkomponente	27
Vertikalstab	105
Vertragsabschluß	6
Verwerfung	43
Verzinkung	119
V-Fuge	39, 119
Vierkantstahl	101
Visiergerüst	47
V-Naht	104

Vogelperspektive	9
Vollgatter	118
Vollwandträger	105
Voranstrich	166, 167
Vorarbeiter	34
Vorentwurf	5, 6
Vorgänge, rechnerische	21
Vorhangschiene	130
Vorhängeschloß	160
Vorkittung	158
Vorlack	166
Vormauerung	89
Vorratsraum	15
Vorschlag	11
Vorschlaghammer	84
Vorspritzwurf	92
Voutenbalken	62
Voutenschräge	62

W

Wabenträger	105
Wärmeisolierung	140
Walm	67, 125
Walmdach	65
Walze	49
Walzglas	157
Wand, doppelschalige	89
Wand, nichttragende	61
Wand, tragende	61
Wand unbelastete	61
Wand, verputzte	13
Wandanschlußblech	109
Wandauflagerholz	126
Wanddurchbruch	40
Wandfliese	13, 94, 163
Wandfundament	14
Wandhängeschrank	122
Wandhahn	150
Wandkasten	123
Wandlatte	126
Wandleuchte	144
Wandplatte	95
Wandscheibe	64
Wandschlitz	40
Wandschüssel	152
Wandsockel	68
Wandteppich	163
Wannenausbildung	141
Warmwasserbereiter	151
Warmwassergerät	145
Warmwasserleitungsnetz	146
Warmwasserspeichergerät	146
Warmwasserversorgung	146
Warmwasserzirkulation	146
Waschbecken	151
Waschtisch	151
Wasseranschluß, provisorischer	48
Wasserbehälter	146
Wassergewinnung	146
Wasserhahn	150
Wasserhaltung	54
Wasseruhr	146
Wasserwaage	84
Wasserzementfaktor	74
WC-Brille	152
WC-Deckel	152
WC-Zelle	45
Wechsel	125
Wechselschalter	144
Wechselstrom	42
Weichasphalt	141
Weiche	103
Weichfaserplatte	115
Weide	113
Weite, lichte	28
Wellasbestplatte	138
Wellblech	108, 138
Wellblech, gekantetes	138
Welle	41
Wellental	41
Wellenberg	41
Wendeflügel	128
Wendeltreppe	66
Werkplan	6
Werkstatt	15
Wetterbrett	67
Wetterschenkel	134
Widerlager	28
Widerstandsmoment	23
Windlast	25
Windstrebe	105
Winkelblech	108
Winkelstahl	101
Winkelstahl-Pfette	106
Winkelzulage	77
Wölbnaht	104
Wohnfläche	7
Wohngebiet	7
Wohnhaus	4
Wohnraum	15
Wohnsiedlung	7
Wohnsiedlungsplan	7
Wohnung	4
Wohnzimmer	15
Würfel	22
Würfelmuster	162

Y

Y-Achse	101

Z

Zähler	143
Zählerkasten	90
Zählerraum	15
Zange	99, 117
Zapfen, Zapfenblech	116, 119
Zaun	14
Zeichenpapiere	12
Zeichner, technischer	11
Zeichnung	5
Zeichnungskopien	12
Zeltdach	65

Zementestrich	93, 162
Zementfabrik	17
Zementlagerung	72
Zementmörtel	75, 86
Zementpflaster	93
Zementputz	92
Zentrierstift	95
Zerrbalken	55, 62
Ziegeldecker	137
Ziegelmaurer	35
Ziegelpflaster	93
Ziegelstein	37
Ziegelwand	13
Ziermauer	61
Zimmer	13
Zimmermann	35
Zimmermannsarbeit	36, 112
Zimmermannshammer	118
Zink	98
Zinkchrom-Mennige	167
Zinn	98
Zinnblech	108, 138
Z-Profil	101
Z-Stahl	101
Zug	27
Zugangstürchen	153
Zugband	126
Zugbewehrung	76
Zugschalter	144
Zugschnur	143
Zugstab	105
Zuschlagstoffe, feine	72
Zuschlagstoffe, grobe	72
Zweispännerhaus	4
Zwinge	117
Zwischenboden	123
Zwischengeschoß	14, 17
Zwischenseite	123
Zylinder	22
Zylinderschloß	160

Fachwörterbücher
aus dem Bauverlag
Architektur · Bautechnik · Baumaschinen · Baustoffe

BAUVERLAG

Wörterbuch für Bautechnik und Baumaschinen
Von H. Bucksch. Format 12,5 x 17 cm.

Englische Ausgabe:
Dictionary of Civil Engineering and Construction Machinery and Equipment
Band 1: Deutsch-Englisch. 8. Auflage. 1184 Seiten. Rund 68 000 Stichwörter. Plastik DM 180,—
Band 2: Englisch-Deutsch. 8. Auflage. 1219 Seiten. Rund 71 000 Stichwörter. Plastik DM 180,—

Wörterbuch für Architektur, Hochbau und Baustoffe
Von H. Bucksch. Format 13,5 x 20,5 cm. Plastik.

Englische Ausgabe:
Dictionary of Architecture, Building Construction and Materials
Band 1: Deutsch-Englisch. 2. Auflage. 942 Seiten. Rund 65 000 Stichwörter. DM 240,—
Band 2: Englisch-Deutsch. 1137 Seiten. Rund 75 000 Stichwörter. DM 240,—

Illustrerad byggnadsteknisk engelska och tyska
Bautechnisches Schwedisch und Englisch in Bildern
Illustrated Technical Swedish and German for Builders
Von W.K. Killer. 1984. Ca. 185 Seiten mit zahlreichen Abbildungen. Texte dreisprachig in Deutsch, Englisch und Schwedisch. Format 17x24 cm. Kartoniert ca. DM 34,—

Wörterbuch für Metallurgie, Mineralogie, Geologie, Bergbau und die Ölindustrie
Englisch-Französisch-Deutsch-Italienisch
International Dictionary of Metallurgy, Mineralogy, Geology and the Mining and Oil Industries
English-French-German-Italian
Zusammengestellt von A. Cagnacci-Schwicker. 1530 Seiten. Rund 27 000 Stichwörter. Format 15,5 x 23,5 cm. Gebunden DM 110,—

Englisch für Baufachleute
L'anglais dans le bâtiment
Von Prof. Dipl.-Ing. G. Wallnig und H. Evered F.C.S.I. Format 17 x 24 cm. Kartoniert.
Band 1: 6. Auflage 1978. 101 Seiten mit 35 Abbildungen. DM 16,—
Band 2: 2., durchgesehene Auflage 1977. VIII, 192 Seiten mit Abbildungen. DM 38,—
Band 3: 1984. Ca. 250 Seiten mit Abbildungen. Ca. DM 45,—

BUCKSCH-REPRINT
Wörterbuch Bau. Ingenieurbau und Baumaschinen
Building Dictionary. Civil Engineering and Construction Machinery and Equipment
Von H. Bucksch. Deutsch-Englisch/Englisch-Deutsch in einem Band. Unveränderter Nachdruck der 3. Auflage. 934 Seiten. Format 12 x 17 cm. Kartoniert DM 38,—

Getriebe-Wörterbuch
Dictionary of Mechanisms
Von H. Bucksch. Deutsch-Englisch/Englisch-Deutsch in einem Band. 286 Seiten mit zahlreichen Zeichnungen. Zusammen rund 16 000 Stichwörter. Format 13,5 x 20,5 cm. Plastik DM 165,—

Holz-Wörterbuch
Dictionary of Wood and Woodworking Practice
Von H. Bucksch. Format 13,5 x 20,5 cm. Plastik.
Band 1: Deutsch-Englisch. 2. Auflage. 461 Seiten. Rund 20 000 Stichwörter. DM 85,—
Band 2: Englisch-Deutsch. 536 Seiten. Rund 20 000 Stichwörter. DM 95,—

Gips-Wörterbuch
Gypsum and Plaster Dictionary
Dictionnaire du gypse et du plâtre
Deutsch-Englisch-Französisch
Von Dipl.-Ing. K.-H. Volkart. 1. Auflage. 176 Seiten. Rund 3000 Stichwörter. Format 17x24 cm. Gebunden DM 85,—

Bauverlag GmbH · Wiesbaden und Berlin